RIVERSIDE COMMUNITY
1916

D0211077

ONE SUNSET A WEEK

BY THE SAME AUTHOR

Joy in Mudville

ONE SUNSET A WEEK

The Story of a Coal Miner

GEORGE VECSEY

SATURDAY REVIEW PRESS | E. P. DUTTON & CO., INC. | NEW YORK

Library of Congress Cataloging in Publication Data

Vecsey, George.
One sunset a week.

1. Coal-miners—West Virginia. I. Title.
HD8039.M62U6698 331.7'62'23309754 73-22484

FIRST EDITION

10 9 8 7 6 5 4 3 2 1

Published simultaneously in Canada
by Clarke, Irwin & Company Limited, Toronto and Vancouver
ISBN: 0-8415-0320-6

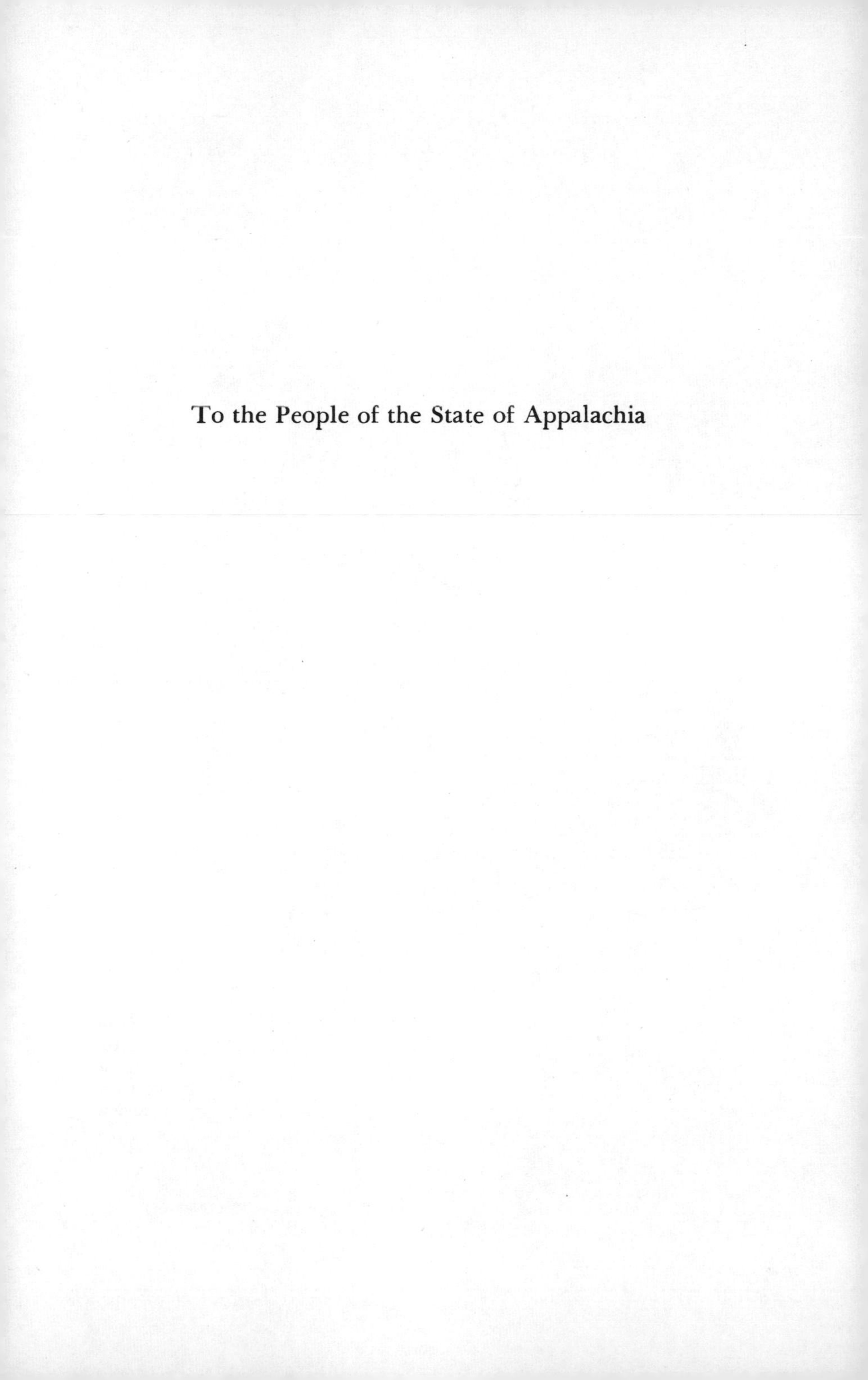

To the People of the State of Appalachia

Prologue

In 1970, after a decade of sportswriting, I went to cover Appalachia for *The New York Times.* While reporting a strike by the United Mine Workers, I heard about a wise old miner who might explain the coal region for me.

After a long ride through the tortuous hills of Southwest Virginia, I reached the Sizemore home at the end of a dirt road. Dan and Margaret Sizemore were waiting on their steep front porch. Quietly, the woman with the dark red hair and the soft intelligent eyes began fixing chicken-noodle soup and egg-salad sandwiches, letting her husband do all the talking, at first.

"You've come to write about this poo-er, poo-er land of Appalachia," he said, broadening his mournful mountain twang, conscious of his effect. "Well, it's a terrible sad story to tell. We look like hell today—but we'll be worse tomorrow and the next day."

As he described how miners were endangered by their companies, by their union, and by their government, I realized that Dan Sizemore was no ordinary coal miner—not with the hard-covered books covering one wall, not with his despair about "the system."

But as a man who had been moving coal for thirty-six years—starting as a willing bully for the owners, becoming a bitter critic of private ownership—he seemed like a valuable closed-circuit camera, planted in one of the darkest, dirtiest tunnels of American life.

So I came back many times—first as a reporter, later as a friend, now as one who loves the Sizemores as family.

Dan and Margaret also helped me appreciate all the other fine people and haunting mountains I came to love. As my teachers, the Sizemores made me believe that millions of people would be better off if coal-bearing portions of Virginia, Tennessee, Kentucky, Ohio, and West Virginia were united into one state of Appalachia—with the power and the profit staying home.

In short, the Sizemores gave me a second home. I visit them when I can. Other times, back home in the hills of Long Island, I can close my eyes and return to their living room, feeling their strength and love.

I hope it is apparent that Dan and Margaret Sizemore and their eight children are real people. Only the names, as the saying goes, have been changed. For reasons that will become obvious.

George Vecsey

Port Washington, New York
January, 1974

Monday

1

Dan Sizemore wakes up coughing. His cough is deep and dry, not something transient like a heavy chest cold or even pneumonia—something you can cure—but permanent, at least as permanent as breathing itself.

His first motion is to reach for a cigarette and start smoking it just as fast as he can. The nicotine is wonderful as it starts to hop up Dan's nerve endings. And the best part about smoking this early in the day is that it takes the pressure off Dan for the rest of the day. Having already lost the battle, he doesn't have to worry about fighting temptation.

Like most of his buddies at the mine, Dan smokes a couple of packs a day. He has a theory about why he smokes so much: if he inhales enough of that wonderful smoke, by the time he gets into the mine this afternoon, right up against that working face with all the coal dust billowing around his nose and mouth, why, buddy, he'll have no reason to fear that coal dust.

Sunday poses a problem, since the mines don't run that day. But Dan tries his best to smoke an extra pack on Sunday, so as not to throw his system off.

Sitting on the edge of the bed, Dan knows it must be around ten o'clock. That's the time the sun rises over the steep ridge behind the house, burning away the mist that lingers in the hollow. The beam of light approaches the barrel of the rifle propped in the corner. Dan still keeps the rifle in his bedroom, even though it is two years since that fearful summer when he and Margaret sat by the window, waiting for the gang of men to come again.

The other half of the bed has been empty since six o'clock, when Margaret got up to prepare for her day at college. Dan still feels a twinge at waking without Margaret in the house. After more than twenty-five years of marriage, he had grown used to her being downstairs, with his first cup of coffee waiting.

The rest of the house is quiet, too. Five of the eight children are off to school for the day—Mary-Ann to college, Vicky and Bobby to high school, Gene and Edward to the wretched little grade school.

Only David is home—probably rocking on the porch downstairs, waiting nervously for Dan to awaken. The principal told David he couldn't finish the eighth grade this year because he was already nineteen years old. It was a dreadful blow to the sunny child who had looked forward to graduation like his older brothers and sisters. Now he stays home every day because there are no training facilities for even mildly retarded children, not in Bradshaw County, in this isolated corner of Southwest Virginia.

Two other children are not home. Pete and Chris, the oldest. They will not be coming home again, not unless that hard-faced man in the White House changes his mind about amnesty for draft-dodgers. Even so, it would not be safe for the boys to come home from Toronto. Too many local patriots, who "did their duty" in Vietnam, would love to bushwack the two oldest Sizemore boys.

Dan knows something about being ambushed. It was only two years ago that the deputy sheriff pistol-whipped him on the lonely ridge road leading to the mine.

Dan stands up tentatively, pads over toward the window, his bare feet standing in the sunlight. He looks more owlish than usual, with his yellow hair rumpled and his dark-rimmed glasses not quite straight.

He peers up the hill, like a ship captain checking the clouds and the wind. Ever since the ridge was strip-mined three years ago, every rainstorm has brought disaster, washing mud and reddish coal slag ("red dog") into the backyard. Dan's back aches when he even thinks about the days spent raking red dog.

But there is no storm brewing on this sunny late-October day. The home is safe for a while longer.

The house is built on the side of a steep, grassy hill. The slag heap and the ridge loom over the rear of the house. But the front of the house looks over a narrow valley, with a creek running alongside the two-lane highway that links Southeast Kentucky with Northeast Tennessee, over this final sliver of Virginia. Only two houses sit on the same plane as the Sizemore home. Down in the valley is a traditional coal camp—identical wooden buildings strung out alongside the road.

Across the valley runs the opposite ridge, thus far spared from the bulldozers of the strip miners. The ruined ridge and the spared ridge wander southward until they merge in Dan's vision. The sensation, even from this second-story window, is of being enveloped in folds of mountains.

Dan likes the feeling of being tucked near the bottom of valleys, sheltered by tall ridges. He never feels uncomfortable when the Sizemores visit their sons in Toronto, walking on the sidewalks of the city, the skyscrapers acting as surrogate mountains for the hill family.

Yet he feels exposed, endangered, desolate, when he drives home along the flatlands of Michigan and Ohio, anxious for the weathered old Appalachian chain. He cannot help but shiver for joy when he spots the first foothills of Southeast Ohio, or when he crosses the river into West Virginia, his native state, heading back to this adopted home.

The view this morning is bleak, with drab rocks visible on the hillside and the coal camp quite noticeable through the trees. Only a few weeks ago the valley was vivid with autumn colors. A few months ago it was so lush and green that you couldn't see the face of the opposite ridge. And a few months before that the early buds exploded after the spring rains. But now the ridge is ready for winter.

To the outsider, late fall is a bleak season in the mountains, a time of hopelessness and hibernation, when all of Appalachia's wounds are exposed. Even to the miners, late fall is a reminder of frigid air drying out the mine shafts, circulating the dreaded coal dust that can explode with the slightest spark.

But not even the encroaching cold or the barren view can depress Dan Sizemore. Winter is his season. He feels challenged by its rawness. He thrills to the desperation of a bitter winter moving in. He is fifty-five years old now. After thirty-six years of working in the mines, his tastes have been tempered by his surroundings. He enjoys starkness, even darkness. While other people talk of building homes on sunny hillsides, with broad picture windows and patios, Dan thinks that his dream house would be Spanish in style, with inner courtyards and thick shutters to keep out the sunlight. And in late October, with the colors gone from the trees, Dan still delights in crisp clean overalls and heavy lunch pails and the challenge of moving the coal.

Not that Dan feels any particular loyalty to his company or his industry. It has been almost fifteen years since the Franklin

Coal Company laid him off so heartlessly, right in the middle of Ike's Recession.

Dan had nine long months to think about the coal industry, nine months out of work when he questioned so many things about America and the profit system that he became a radical, an outsider, to most people in the coal community. Yet even though he believes that his current employer, the Big Ridge Coal Company has lost its ability to run a mine, he still cannot imagine doing anything else on this fall day than going to the mine. He has become a work beast, just like his daddy was.

Dan still stands in the window, in no hurry to go anywhere. He can see only two houses below his, at the end of the dirt road. Next door, in the old Montgomery house, Dan can see the new woman hanging drapes. This must be the tenth family to live in the Montgomery house since poor fat Lester Montgomery committed suicide there, two hot summers ago. Some say there is a jinx that destroys families that live there. Others say the smell of Lester Montgomery's body, lying for days in his room, will never go away. Dan tries not to think about helping to carry Lester's body out of the house.

Dan has not met the new people, although he saw them on the back porch two weeks ago, in a brief episode that has kept the Sizemores laughing ever since.

It was a rainy Sunday and the Sizemores were lounging around when they heard this tremendous shotgun blast across the grove. They looked out the window and saw the new man standing on his protected porch, holding his smoking shotgun. With violence always on their minds, the Sizemores immediately feared the worst when they saw the woman running through the tall wet grass, holding her baby in her arms. But then she reached down and grabbed the body of a huge gray squirrel. Waving it proudly, she rushed back to the porch to present it to her husband.

Dan told Margaret and the two girls, "Instead of going off to college or reading those women's liberation magazines from

New York, that's what you-all should be doing." He giggled loudly when they took the bait.

Still smiling about the trio of sensitive women in his family, trying to get along in the bowels of the coalfields, Dan walks slowly toward the stairs. Coughing softly, more from the coal dust imbedded forever in his lungs than from cigarettes, he begins the strenuous business of walking downstairs.

In the kitchen Dan makes a cup of instant coffee from the kettle that works twenty-four hours a day, the acidulous mountain water turning the inside of the pot a deep rust color.

Dan takes his cup into the living room and sits in his favorite chair, next to the bookshelf. Within his reach are books by John Steinbeck and Philip Roth, Eldridge Cleaver and Karl Marx, held upright by the ends of shoe boxes.

Over by the phonograph the albums are scattered on the floor: Three Dog Night, Joan Baez, Loretta Lynn, plus an album by a local musician protesting strip mining, entitled quite accurately *You Can't Put It Back.*

Lying at Dan's feet are the newspapers and magazines he reads: clippings from *The New York Times,* sent by a friend up north; the *Whitesburg Mountain Eagle* ("It Screams"), a crusading weekly paper from Kentucky; and *Coal Patrol,* a muckraking coal monthly from Washington.

On the bare wooden coffee table is a glossy hard-cover book about Communist China. Dan has great admiration for that country. Everybody works; everybody is taken care of; nobody makes a profit, Dan thinks.

Dan picks up the morning paper that David has brought up from the delivery box on the main road. The paper is the *Tri-Cities Chronicle,* serving the towns of Bristol, Kingsport and Johnson City, where Tennessee and Virginia come together. Dan believes the paper slants the news in favor of Nixon and Agnew and all their brutal-looking henchmen, in favor of the coal operators, in favor of the flag-waving congressmen from the area.

The national news is taken from the wire services, edited down until it is meaningless. Dan often does not understand a complicated political or financial story until he reads a weekly magazine or somebody sends him a clipping, weeks later. He exists on letters from friends, on long-distance telephone calls, on news that has become calcified into history, into books.

This morning Dan reads a story about a wildcat strike in a neighboring mine. Dan has been a foreman, a company man, all his life. But his loyalties are toward the miners who overturned their water pails, the traditional signal for a wildcat strike.

"This goddamn paper makes it seem like they strangled their mothers," Dan hisses, his soft drawl gone vicious for a moment. "I wonder how much the coal operators pay for this kind of story."

But he reads on. His mind is clear in the morning. He is hungry for news. When he finishes the paper, he starts on the magazines. The time passes.

At 1:30 Dan begins packing his tin lunch pail, a skill he has acquired late in life. Margaret used to fix his lunch every day, stuffing his pail full of thick sandwiches and hot soups and cookies. But now that she is a college girl she is rarely home on a weekday.

Deep down inside, Dan is proud of Margaret going back to school after twenty-five years and making the dean's list. But that doesn't mean he can't grumble once in a while about making his own lunch.

"Oh, I almost forgot the cheese," Dan says, returning to the refrigerator for the last wedge of provolone cheese I brought down from New York.

"You should see all my buddies on the shift," Dan says. "They won't leave me alone until I share this. It makes me realize how bland our food is down here. God, when we visit Pete and Chris up in Toronto, I can't get enough of the different kinds of foods."

He goes upstairs and dresses to leave. He will spend much of the evening in his dirty work clothes, wallowing in dust and grease. But for the ride over to the mine he wears a freshly pressed sport shirt and clean slacks, his hair slicked down, his face smelling of soap and shaving lotion. Dan says that miners spend so much time being dirty that they like being double-clean at home.

At 2 P.M. he drinks his last cup of coffee and eats another piece of toast. Next he takes a flashlight off the counter and tucks it into his lunch pail. Then he strides over to the mantelpiece and takes a pistol down. He stuffs the pistol into his pants pocket. He has been carrying the pistol ever since the deputy whipped him, two years ago. Dan says that one of these days he is going to start leaving the pistol home again.

But not just yet.

2

Usually Dan walks down the dirt path and takes a ride to work from one of the miners in the camp below. But today he will drive with me.

Dan's 1968 Ford is parked in the dirt clearing below the steep porch stairs. He bought the car secondhand from one of the miners last year. The Sizemores had previously gone without a car for three years, to economize while they sent money to Pete and Chris in Toronto.

His car is modest for a foreman who makes more than the top union scale of nearly fifty dollars a day. But with eight children, the money just seems to go.

Dan gazes longingly at the rental car, a shiny Galaxie with only a thousand miles on it. We get in. The house looms above us. The ridge looms up behind the house. Backing up slowly, the car is maneuvered around several rocks and trees. Then forward.

The car creaks as it inches slowly down the path, past several houses and a garden planted in a clearing. Dan says he must find time to repair some of the worst ruts. The car tires come within three inches of the steep edge of the hill, where the rain has washed away a precious portion. At the bottom of the path we make a sharp right turn, going only a few miles per hour, past a small wooden house where a woman is hanging clothes on the porch. She stares impassively at us, her face saying, "Oh, there goes that Dan Sizemore with another of his long-haired hippie visitors. There's always some outsider visiting those people."

We turn left and begin climbing the main highway toward Milan, past the familiar Appalachian landmarks that visitors notice first—the ramshackle gas stations, the tiny groceries with signs advertising R.C. Cola and white bread, the junked cars piled at the side of the road.

Dan motions for me to stop at a row of identical houses, where old people and little children sit on the porches in the afternoon sun. He shouts, "Hey, would you tell Joe I won't be riding with him today? I got me a ride already."

The old man on the porch nods and relays the message through the doorway. We drive on. The house belongs to Joe Brown, one of the night-shift miners over at the mine. Joe is a friend of Deputy Sheriff Clyde Turner, who bashed in the side of Dan's head with a pistol handle two years ago while Joe was driving over to the mine. Joe professed he knew nothing about his friend's intentions that day. They dropped the subject a long time ago.

The hill gets steeper as we climb Milan Mountain. The road has been dug into the hillside, leaving barely enough room for two lanes. We are on the outside lane. There is no railing to keep a car from tumbling five hundred or a thousand feet down. Three times the road takes a sharp left turn, where the driver cannot see what is bearing down on him. One must always be alert for the massive coal trucks, dangerously overloaded, careening downhill, taking their half of the road from the middle. The drivers are paid by the number of loads they carry to the loading station. When they run somebody off the road, they have been known not to look back.

Even when the coal trucks are not barreling downhill, the mountain throws little surprises at the driver, like rocks of eight ounces or five hundred pounds after a rainstorm loosens up the surface. Or sometimes a driver in the opposite lane will make a sudden swerve to avoid hitting a recent rockslide.

Then we reach the top of the ridge. The view is spectacular on this clear day, the ancient mountain ranges stretching horizontally—not peaked like the newer Rockies but weathered down, rippling like the backbones of a giant whale herd. Kentucky's dense convoluted mountains lie to the north; Tennessee's long ridges and gentle valleys flow southward of this sliver of Southwest Virginia.

"They call this section 'the Top of the World,' " Dan says. "The people up here are ridge people. They're not coal-camp people like us. Some of them talk a dialect you could hardly understand—like ducks. Kwa-kwa-kwa, through their noses. They like living on the ridge. They're a different breed."

Some of the ridge buildings are modern brick ranch homes with new cars parked outside. They have grape arbors, honey boxes, some horses and cows grazing in the fields. The ridge seems relaxed and sunny after the tight, shadowy feeling of the valley. Other ridge homes are wretched shacks, with outhouses and rusted junk cars.

Suddenly the pavement ends and we are driving on hard-packed dirt, rutted and bumpy.

"They used to scrape this road," Dan says, his voice rising into a nasal whine. "But then old John Drago, the supervisor over in Milan, bought a helicopter for his monthly visit to the mine. Now they don't scrape the road anymore. The forty-nine vice-presidents all drive company cars, so they don't give a shit what happens.

"And the poor miserable little miner, who ruins his own automobile going down this horrible road? Why, it's simple. He gets his revenge by destroying half a million dollars' worth of mining equipment one night when the foreman ain't looking. But the company doesn't understand that. Drago takes his orders from the accountants in New York. He stopped listening to his foremen years ago."

The road is descending at a fearful angle now, fifteen or twenty degrees, the car tires grabbing at anything they can touch—rocks, mud, pieces of coal, pieces of steel, rubble fallen off cars or trucks. The feeling is of minimal control, like an airplane trying to land during a thunderstorm, just skittering back and forth, trying to get a grip.

"It's worse in the winter," Dan says. "Some nights it snows and then ices up and we just slide all the way down. It's a fearful thing. Guys wreck up all the time. It's a wonder we haven't lost anybody yet."

We are entering another valley, completely turned over to coal. The upper ridges have been gouged by strip mining. Each narrow hollow has a rocky slide area in the lowest fold, where the boulders and timber tumble down from the strip ledges. The rest of the hill is the thick unattractive green of weeds and fir trees. We are entering a free-fire zone of the soul, a Vietnam of industry. The dust from the road filters into the closed car, makes me choke even worse than Dan's cigarettes.

Yet amazingly, in the midst of this jumble, some gay little birds flutter and swoop: a tiny blue bird sitting on a telephone wire; a little yellow bird, another warbler, darting into the trees. Redbirds—cardinals—as rugged as any coal miner, seem to thrive in this testing ground.

We reach the bottomland, pass the tipple that towers a hundred feet in the air, where coal is dumped in coal cars. The freight engines then pull the cars—seventy to one hundred tons of coal in each of a hundred cars—through a cement tunnel. In ten minutes these cars will pass within a quarter of a mile of Dan Sizemore's house.

"You'll notice the fine tunnels we have for the coal," Dan says bitingly. "It took us forty-five minutes to cross that ridge. But the coal goes out in ten minutes. Tells you where we stand."

One more hill to climb. Dan's mine is located on a ledge a thousand feet in the air. The mine goes straight into the mountain—a "drift" mine. Big Ridge has been digging coal here since 1958. They say there is another dozen years of coal inside. But Dan says the way John Drago is letting the mine run down, they regularly lose 60 percent of the coal. Sometimes whole sections just fall in. Dan says it would not surprise him to see Big Ridge's accountants close down Mine No. 7.

We park in the dusty lot, alongside dozens of Fords and pickup trucks and Volkswagens. Dan walks toward a long tin-roofed building, the bathhouse, a "luxury" the industry installed around ten years ago, rather than raise pensions or improve safety conditions.

In the bathhouse a few night-shift miners are getting dressed. Their work clothes are suspended from the roof in baskets, by individual pulleys. Dan lowers his basket and neatly folds his clean slacks and shirt. Then he starts dressing for work.

A striped denim shirt. Striped denim overalls. Rugged high-cut shoes, with a steel plate built into the toe. Then he runs a piece of adhesive tape around the floppy lower part of his pants leg, to keep excess material from catching in a piece of machinery.

Out of the basket comes the white construction cap that will absorb the stress from any small chunk of falling rock. It used to be that only the bosses wore the shiny white helmets, but lately many workers have been using them as well.

Then Dan straps his broad safety belt around his waist—not to hold up his overalls but maybe to save his life. On the safety belt he will attach his self-rescuer, the precious breathing device that will give him one hour of air in case of an accident underground. The safety belt also holds a battery that powers the lamp he will snap onto the front of his cap.

Dan stands up, looking clumpy and graceless in the bulky outfit. But he has been wearing the uniform for thirty-six years and feels perfectly comfortable in it. He raises the basket toward the roof. Then he strides out toward the yard, where the night shift is beginning to accumulate.

Three o'clock. The men will not enter the mine until the four-o'clock siren, but already there are at least twenty-five of them lounging around the yard and the lamp room, fully dressed.

Two men in green overalls are pitching horseshoes among the shuttle-car tracks, the corrugated-steel shed, and the highwall of the mountain. The clang of their horseshoes is punctuated by wordless yelps of success or failure.

Several other men in work clothes are slouched against another wall of the shed, sitting on the greasy ground, whittling with knives so sharp they would draw envy from any street gang. The men are whittling discarded mine timber, letting the chips fall in their laps. When one piece of wood is decimated, they pick up another chunk and start whittling. Sitting in the sun, the main thing is to keep their hands and their mouths busy.

One of the men plucks five fingers' worth of Red Man Chewing Tobacco from a paper pouch and stuffs it in his mouth. Another man squirts out what seems like a quart of brackish brown tobacco spit, neatly over his shoulder, between the tracks. They raise their tobacco pouches in silent offering, laughingly knowing that the visitor will not dare to try any.

The conversation at the moment concerns Joe Norton, a

former deep miner like themselves who has become a millionaire from his small strip-mining company.

But Norton is not one of those all-work, no-play fellows. The other afternoon he was in bed with some miner's wife when the husband came home early. Normally Norton is never far from his guns—he uses them to scare off the strip-mine protestors—but at that particular moment he was much too involved to reach for his guns when the husband broke into the bedroom. The men hear that their anonymous fellow miner kicked Norton's balls over and over again, sending Norton to the hospital.

"Guess old Norton will have to concentrate on strip mining for a while," one of the miners says without expression. The men are not enchanted to think of Norton making a million dollars stripping coal in the sunlight while they struggle like moles under the mountain.

The men shift the subject to a woman in a nearby coal camp who has boyfriends on all three shifts. They marvel at her indestructibility and speculate how she handles Sundays and holidays. One miner says that if she were his woman, she wouldn't have enough energy for the other two shifts.

"We have some of the world's greatest lovers at the mine," Dan Sizemore whispers. "It is a well-known fact that miners talk sex at the coal mine and they talk mining as soon as they have a few beers at the local bar."

In fact, Dan says, if the indestructible coal-camp woman were to visit this mine at this moment (women are regarded as bad luck inside a mine but not necessarily in the yards), the men would immediately begin telling her what "machine dogs" and "roof-bolting daddies" they are.

While the one group of miners whittles outside, another group waits in the lamp room, sprawled on scarred wooden benches, surrounded by racks of self-rescuing devices and gas masks and cap lamps, all of them maintained by Jimmy Davis, the lamp man. In the lamp room there are two soda machines, a coffee machine, and a bulletin board cluttered with company and union notices. Most of the notices have harsh comments scribbled on them.

Tacked on the board are a series of "Fatalgrams" issued by the U.S. Bureau of Mines, the federal agency that supervises coal mines. According to the Bureau, the ninety-seventh fatality of the year, to a shuttle-car operator in Shinnston, West Virginia, was caused by "operator's failure to observe buckled rib near crosscut."

Underneath, one of Big Ridge's miners has written: "The company is never at fault. Pigshit."

The lamp-room group keeps growing as the clock sweeps toward 3:30. Most of the miners are in their forties and fifties, with the rough, worn faces of men who have spent much of their lives doing brute work. But there is something gentle about the way they clear room for a latecomer and offer to buy a soda for the visitor. A few of the miners are in their sixties, their eyes clouded with visions of retirement or death. And sprinkled among them, every so often, is a younger man with mustache and long hair, his eyes also distant, from rock music perhaps, or from his year in Vietnam. The middle-aged men dominate the conversation in the room.

They have been debating whether the mine will be shut down soon. The company has been sending some cold-eyed investigators over, every couple of weeks. The way John Drago is running the mine into the ground, the gloomier men say they would not be surprised if Big Ridge just sealed off the mine mouth and wrote it off. Other men say, "Hell, we'll be working here when we're eighty."

Right now the speaker is Raymond Wilcox, who has a flat crew haircut, right out of the 1950s, when he was a young man starting to mine coal.

"I started out in 1956 and moved over to Big Ridge when we opened in 1958," he says. "We used to work without lunch if the bosses felt like it, but that doesn't go over anymore. First time I got hurt was 1959. I got covered by a rib roll [a collapsing wall] and got three vertebrae busted and a busted pelvis. I got a 25 percent disability for that, but I came back to work anyway.

"In 1965 or '66, I forget which, I slipped in some grease and broke my neck and got my back messed up. It was my fault. I was in too much of a rush.

"One year I had a stomach operation and the doctors pronounced me totally disabled. But I came back and worked four more years. In 1971 I hurt my right leg and right arm and was paralyzed."

Why did he come back?

"I don't know," he says, looking around with his brown bird's-eye glance. "The company's been good to me in a way. I kind of hate to sit around. On the other hand, you got to produce.

"Oh, yeah. I've also got first-stage rock dust—silicosis. I got paid a flat $3,100, but my lungs are so bad I can't hardly get around."

His buddies half-listen to Raymond's story. Most of them can match it, fracture by fracture, wheeze by wheeze. Although it is not apparent from the way they move in their clumpy mining gear, which makes everybody seem slow, many of them have some physical disability. Seen in street clothes, a lot of them walk in the slightly disjointed fashion of something that has been broken and put back together again, like marionettes.

Particularly the hands. Shaking hands with a coal miner is always an adventure. He slips you his hand and you sort out the damage without showing any expression. Hmm. Top joint missing on index finger. Hmm. No thumb. Hmm. Two smallest fingers gone. Hmm.

And there is no sign that mining is getting any safer. The American coal industry has produced over 100,000 deaths and 1 million injuries since 1900—and the men know that fatality rates are at least twice as high in America as in the European mines. There were 132 fatalities in the United States in 1973, amounting to 0.45 deaths per 1 million man hours—the most dangerous occupation in the country, contrasting with the national manufacturing rate of 0.03.

The fatality rate has dropped each year since 1970, the first year under the Coal Health and Safety Act. In 1970, 260 men died at a rate of 1.00 deaths per 1 million man hours.

However, the drop in deaths may be a result of better medical care rather than mine safety—because injuries have stayed at 44 per 1 million man hours according to the Bureau of Mines. The men generally do not trust material from the Bureau of Mines, which has been dominated by complacent academics, political fund-raisers, and unsophisticated young hustlers over the last three years.

Traditionally the men don't get much protection from the union, either. Many of them feel the union has neglected them in matters of health and safety. In 1969 they saw a challenger for the union presidency, Joseph A. (Jock) Yablonski, gunned down in his bed. Later they saw the president of the union, W. A. (Tony) Boyle, convicted of misusing union funds and other sins.

In 1972, after unprecedented court and government involvement, the miners voted themselves a new president, a soft-spoken but rugged miner, Arnold Ray Miller. And in 1973 the men saw Tony Boyle indicted for complicity in the murder of Jock Yablonski and they saw Boyle attempt suicide on the day before the extradition hearing in the murder charge.

The men innately trust Arnold Miller, who has visited their mine, but they still consider themselves cut off from all sources of power.

The miners feel themselves directly at the mercy of the Big Ridge headquarters in Milan—with no immediate recourse to any other authority. When they feel the company is abusing them—pushing them toward higher production—they feel their basic weapon is a wildcat strike, even though Boyle never backed them up and Miller urges them to comply militantly with the proper grievance procedures. But the men walk out anyway, when the company threatens to bump a senior miner from a good job or his regular shift, even though a day's walkout will cost them forty-odd dollars in salary.

"Heck, the money ain't worth that much anyway," says one miner. "When Boyle got us that new contract in 1971, the grocery store raised its prices twice—once when it heard about the raise and once when we got it. I'm worse off now than I was then. I save money by staying home."

"Last time you stayed home you got yourself a new baby," one of his buddies says dryly.

The men are considering a walkout next week over some job grievance. It would come at the right time, the opening of deer-hunting season. There are not many deer left in the hills, but the men will hunt on any pretext, particularly since it keeps them out of the mine. It is one of the paradoxes of mining that the men come an hour early to work each day but stay away completely whenever they can afford it.

"It used to be Hangover Monday," Dan Sizemore says. "But now it's Hangover Tuesday, Wednesday, Thursday, and Friday too."

If the men are told to work on Saturday, they grumble, even though it means time and a half. Making close to $150 for three days' work is not a starvation salary in the mountains. Most of these union miners own campers and color television sets and good kitchen utilities for their wives. A day out of the mine is another day saved for their blackened lungs.

The lamp room starts to fill up as the big wall clock works its way to 3:45.

A slender man in immaculate green denim work clothes slips his way inside the door, like a garter snake, silent and trim. He edges along the wall, sneaking behind a big bear of a miner sitting on a bench. The slim man jabs several of his fingers onto the burly miner's rear end, causing him to jump off the seat.

"Whoooo!" the big man exclaims.

The skinny man glides out the door again, grinning.

"That's Dean again," shouts the big man. "That goddamn little Dean. I swear I'm gonna wring his goddamn neck one of these days—even if they do shut us down."

The men roar at his anger. Estill Dean is the joker of the shift, full of surprises, goosing his buddies, drawing caricatures of them, switching lunch pails. Even when he was Tony Boyle's staunchest supporter on this shift, he could get away with the sneakiest of practical jokes to a Yablonski-Miller man, just because he was Dean.

The door opens again and a sturdy black man walks in. He is Calvin Brooks, the only black employed among the 175 men at Big Ridge's three shifts. The company has resisted some minor pressure from black communities to hire more black men, saying it has no jobs available. But Calvin has been working here since the mine opened. He carries himself with the casual poise of a man who has proven himself so often that he can now relax and enjoy life.

"That's Calvin Brooks," a miner says to a visitor, loud enough so Calvin can hear. "Calvin doesn't need to work in our mine. He's dark enough already."

Everybody laughs at this remark, even Calvin. He drops a dime in the soda machine and waits until the paper cup is filled. Still chuckling over the joke, he walks outside. The men trade stories about how Calvin can fight, can run a continuous mining machine, can drink anything ever bottled. The men recall the only time they ever saw Calvin turn down a drink—on a Tuesday morning after a three-day weekend. Calvin said he had drunk so much over the holiday that all he could eat that morning was a raw onion. Nobody asked him why a raw onion. They just accepted it as Calvin.

The volume of joking increases as the clock moves toward 4 P.M.

As a former sportswriter, I feel something familiar about the mood in this lamp room. And then it clicks. I can recall the most enjoyable baseball team I ever covered, the Saint Louis Cardinals of the middle 1960s, with their great depth of character and skill. I can remember them teasing each other, Bob Gibson teasing Roger Maris, Timmy McCarver teasing Curt Flood, Bill White and Dick Groat huddled in a discussion of hitting techniques, twenty-five different men merging their personalities in a lusty, respectful atmosphere.

Strange to associate this dangerous business with something as joyful as playing baseball. Yet the comparison goes beyond the long wooden benches in the lamp room and the sickly sweet odor of the tobacco juice. In both cases, I have the feeling that these men would rather be together, earning their living and playing pranks, than staying home with their wives and children—or working some dreary office job with more security.

Now, just as the ball players used to clatter in their spiked shoes toward the field, the miners begin to clump toward the rail yard as the clock approaches four.

Outside, the electric railcars are clacketing out of the concrete mouth of the mine, low metal-roofed cars releasing their cargo of day-shift miners, who rush past us without hardly a word or gesture of greeting, their eyes glittering beneath the layer of dust and grease, their minds on their showers and their getaway.

Slowly the men of the night shift pile into the railcars just vacated by the day shift. Each section crew occupies its own car (or "man-trip jeep"). There are five sections working tonight, three with eleven-man crews, two with seven-man crews. The men have varied jobs inside—machine operators, shuttle-car drivers, roof-bolters, repairmen, drillers, cutting-machine operators, explosives men.

Then smaller man-trips carry the free-roaming extra men—the supply men, the troubleshooters, the belt men, and the supervisors.

As the car operators reverse the electric pole, touching the overhead cable, the miners seem to quiet down.

They stuff a last-minute handful of chewing tobacco or candy bar into their mouths. They secure their lunch pail between their boots. They prepare their heads for the work inside. For all their nonchalance about mining, they know they must be alert inside. A miner who is thinking about something else—about sex or sports or the garden he is growing in his yard—is quite likely to have the roof fall on his head because he didn't pay attention to the creaking and the crumbling overhead. So the miners grow silent.

On their way toward the working face, each crew will stop at its respective "dinner hole," a dugout retreat where they will be instructed by the section foreman. The crew will also be entitled to say a prayer. Some crews pray fervently, others ignore the ritual; it depends on the men. Then they walked the last few hundred yards up to the working face, to run some coal for the stockholders up in New York. The men will be under ground for the next eight hours.

The siren goes off. The first operator shifts the car into gear. The man-trip creaks and rattles toward the mine mouth. The men in the rear cars sit impassively. A few of them smile at me, holding out a pouch of tobacco mockingly or beckoning with their hands.

"Come on with us," says the big miner that Dean goosed back in the lamp room.

And then they are gone.

* * *

Dan Sizemore watches the last car disappear into the mine mouth. It is suddenly very quiet. For the last hour, he has been checking reports from the day shift and making telephone contact with the expediter, the command post for all four Big Ridge mines scattered throughout this valley.

"Quiet day today," Dan says, seeming preoccupied. His doleful wit, his sarcasm at the system have vanished. He seems crisp and loyal and efficient.

For the next eight hours, Dan will be absolutely in charge of this mine, responsible for sixty-seven men now scattering deep inside the shafts, responsible for making decisions that will produce coal and could make the difference between life and death, long before any company official at Milan could be notified. For the next eight hours, it is Dan Sizemore's mine.

The responsibility does not faze Dan. He has been a boss almost his entire working career. There was a time when he thought he was destined to be a company official, like the big boys who visit the mine in the company helicopter. But that was a long time ago, millions of tons of coal ago, and a few bad breaks.

3

Five o'clock. Dan stands alone at a chest-high table, taking notations from the expediter's telephone.

The office is drab and functional, a few tables and file cabinets, dabs of mud and grease and coal dirt everywhere.

Right over Dan's head is a sign saying "SAFETY FIRST."

In a blank corner of the sign somebody has drawn a leering caricature of a miner, with a nasty snaggle-toothed grin. The character is labeled "Dan Sizemore."

Dan is no art detective but he knows the work of Estill Dean when he sees it. That leering face is Dean's trademark. All his subjects come out with the same expression, which (Dan thinks) actually resembles the artist more than the subjects. The miners say that Dean is so ugly that when he was born his mother borrowed another baby to take to the christening.

On the wall next to Dan is a huge map of the mine, six feet by four feet, with red and black shadings noting every single shaft, every pillar, every crosscut, every air vent. The map reminds Dan of a street map of New York that he tried to decipher one time. (Dan and Margaret also tried the subway—once. The train got stuck between stations for five minutes and Dan bellowed, "Why in the goddamn hell are they doing this to me?" He was more terrified of the subway than of any coal mine he ever saw.)

On the huge map the five current working sections are identified by arrows. Dan knows all five sections by heart—knows their strengths and weaknesses, knows the men working on them. He is worried that Ten West is going to fall on somebody's head one of these days. The company is pushing too hard for production in there, not building enough roof support. They can always find a weak foreman on the day shift or the hoot-owl shift who will ignore the basic rules of mining, in return for a raise or a pat on the back. But not on the night shift.

Dan also prepares himself for all that his reports won't tell him—somebody on the hoot-owl shift, resentful at working that ungodly midnight-to-eight shift, did not bother to lubricate the roof-bolting machine. Or somebody forgot to spray a section with the precious rock dust that will keep down the volatile coal dust. Dan knows any of this is possible.

He also knows his common sense is not enough. Some foremen don't trust anybody, but Dan encourages his men to make suggestions about dangerous conditions. Dan didn't use to be that way. But since the layoff by the Franklin mine back in 1958, he decided he was out for himself and his men first. He uses sixty-seven pairs of eyes out in the mine. And he still leads all four Big Ridge mines in this valley in production, because the men work hard for him.

They tease him about his scholarly language; some of them fret about his radical politics; but when they see a rib buckling or their pocket methane gas indicator starts ticking, they know Dan Sizemore will listen to them.

Almost time to go into the mine. But first Dan opens his lunch pail and takes out his first sandwich of the evening. Carefully, before eating it, he pries open the two pieces of bread, just to make sure there is no foreign substance inside.

He got into this habit in recent months as the men tried to trick each other into eating vile-tasting engine grease, so thick and red and shimmering and odorless that anybody would mistake it for red strawberry jelly.

One of the boys set up Deacon Lambert with a grease sandwich just a few weeks ago. When the miners tiptoed up to Deacon's dinner hole, they found the sandwich dashed against the wall—with two huge bites taken out of it. Deacon is now known throughout the mine as the hungry soul who needed two bites to know he was tasting engine grease. Dan would like to avoid that distinction, particularly with Estill Dean on the prowl again.

Poor Dean. The little wretch from Sorrow Creek went into a slump when his hero, Tony Boyle, got whipped by Arnold Miller in the election in December. For a while Dean was concerned that he would lose his minor post with the union. But in the end nobody else wanted to do the dirty work, so Dean remained. Now he is up to his old tricks again, roaming through the mine on the powder-car. Definitely a man to keep an eye on, Dan thinks, chuckling to himself.

The peanut butter and jelly sandwich is untouched by Dean's hands. Dan wolfs it down. Then he drinks a cup of soda and walks out toward the yard.

The mouth of the mine is concrete, for safety and for looks. Dan pilots his small, sputtering jeep through the opening. The rails curve slightly and then he is totally inside, the headlights of the jeep piercing the darkness of the mine.

The shaft is twelve feet high, one of the highest mines in Appalachia, caused by the Samson seam and the Bethel seam running within a few inches of each other. The men of the Big Ridge mine therefore have the comfort of working in natural, erect positions all day long, rather than in the brutal stooping posture of most of their brothers in the coalfields.

But there is a disadvantage to working in a twelve-foot mine. Even a rock weighing a pound or two can put a man in the hospital if it falls from twelve feet. Many miners swear they would rather work in low coal, where they can feel the roof for any signs of a cave-in, rather than rely on their eyes and ears and on sounding devices in a twelve-foot mine.

The cement tunnel lets out after a hundred yards. Now the sides and ceiling revert to the natural rock of the mountain. There is nothing smooth about the tunnel. The rocks are jagged, the floor dips, puddles of cold water lie everywhere, punctuated by heaps of old rock dust and fragments of coal.

The temperature is damp and in the fifties, as always, with a cold wind blowing. In the summer the men cheer when they enter the mine. But sometimes they faint when they leave it, walking into the humid summer air. In the winter it is never warm enough to offer any relief from the snow and cold outside.

Dan feels a good lonely sensation deep in his bowels. He is as isolated as he can be—just himself and this living thing called a coal mine, that sighs, that groans, that creaks, that never rests. The wind and the lamps shining in the puddles create a psychedelic effect, making a man see more than is really there.

The senses work differently inside a mine. Over in West Virginia one time, an invisible man-trip clattered right past Dan's section, scaring the hell out of a dozen men. Dan angrily called up the expediter and asked him why a man-trip was running so close. The expediter insisted that nothing was moving inside that mine, but the men swore on their family Bibles that the man-trip had passed within a few feet of them. They had heard the clatter on the rails; they had felt the tunnel vibrate.

Another time Dan saw red taillights up ahead of him on a deserted section. When he called up the expediter, the same thing happened. The man insisted that nothing had come out of the mine in two hours.

Dan stopped believing in religious mysteries a long time ago. But he believes he experienced ghostly man-trips both times.

The shaft pitches and rolls, following the contour of the coal seam, the vegetation that was pressed together a million years ago. The shaft is hardly as straight and true as a railroad tunnel. But there is little danger of runaway coal cars. Every few thousand feet, a little bug-shaped object between the tracks catches a rung on the underside of the man-trip, automatically preventing it from rolling away. Dan reaches down and manually releases the "De-Rail," then he proceeds.

Also, every few hundred yards, the car glides through a rough burlap curtain, called a "brattice," made mandatory by the 1969 Coal Mine Health and Safety Act. Installed to control the air flow, the brattices are cursed as dangerous by most miners, including Dan. He feels the law was written by men who did not know mining.

Dan reaches a junction where the tracks split in two directions, east and west, the two main portions of the mine. He has traveled half a mile by now, to where they first started digging coal back in 1958.

And this is where the first man died in the first week.

The man was a stranger to Dan. Hell, they were nearly all strangers in those first few months, when the mine was like a Devil's Island or a French Foreign Legion, composed of recruits from all over Appalachia, poor souls out of work in Ike's Recession, ready to move to this lost, lonely corner of Southwest Virginia, just to find work.

Dan doesn't even remember the man's name. All he remembers is how the man started thrashing about, and how the men had called for Dan, the foreman.

"What's the matter with you?" Dan asked.

"What the hell do you think?" the man snarled back.

Dan tried to hold the man down, but his thrashing took them both near the moving coal conveyor belt that was still humming along, moving that coal, threatening to pull them into the metal gears. It took four of them to hold the miner on the ground, before he collapsed. Then they put him on a man-trip and carried him outside the mine, for the hour car ride to the nearest doctor. (Few American mines have ambulances or medical help in attendance.)

The man was dead before he arrived at the doctor. Later they found digitalis pills in his shirt pocket. Apparently he had been a cardiac patient but had concealed it on his new job, for fear of being let go. But he died anyway—nearly taking Dan with him—right there in that side tunnel, now long since mined out and left to the rats and the echoes.

Dan had seen men die before that incident and he has seen men die since. He has seen men squashed into jelly, their inner organs left sticking to the rocky floor. He feels passionately that no man should die in such a dirty place as a mine. He feels there is nothing more wretched than some poor broken man being carried out of the mine, covered with a greasy blanket, with nothing more for his buddies to do than stand around and gawk. Men should die in clean places, Dan thinks.

Now he is more than a mile deep in the mine. As he drives, he inspects the ribs and the roof for cave-ins. He checks the electric cables for poor splicing jobs or signs of wear. He checks for water dripping in dangerous places. He might check for methane gas, although his section foremen are supposed to do that regularly. He watches the belt for signs of stress. Often he is looking for some specific problem, but today he is merely visiting his men, giving them a chance to talk to him.

Now he reaches a working section, a seven-man crew utilizing a continuous mining machine, a marvel of modern technology. The continuous miner is a rectangular vehicle, open on the top, operated by one man. Its front half is lethal, like the jaws of a shark. The top "jaw" has a circular cutting tool at the end that breaks the exposed coal and drops it to the floor. Then the bottom "jaw" scoops up the coal and flips it backward, into the "mouth" of the machine. The coal crunches through the digestive system of this fascinating beast, then is deposited out the back, onto a waiting shuttle car, whose operator drives the coal a few hundred feet to the nearest conveyor belt.

The miner is grinding away as Dan parks his man-trip and walks toward the working face. With every step, the noise and the dust become more unpleasant to Dan's unprotected ears and nose and throat. In very few mines are the workers encouraged to protect their face from the dust. And even then they often refuse to wear masks and earplugs, claiming they would not feel natural and might get hurt.

Better than feeling protected, the miners try to "bid off"—to take advantage of their seniority to duck the jobs at the working face. Machine operators claim they should be paid a higher scale, because of the dust they swallow.

Dan sees Calvin Brooks working the gears of the continuous miner. Calvin waves one gloved hand. Calvin qualified for black-lung benefits several years ago, but he pushes himself to run the machine, convinced nobody could ever replace him.

The other men on the shift nod to Dan. Nobody says much, not with the miner grinding away. Down farther in the shaft, one of the roof-bolters is drilling new holes for the huge bolts that hopefully will keep the roof from falling. Roof-bolters walk into sections that have just been blown open by explosives and shored up with some timber. They lead the league in fatalities and they like to bid off that job, too.

Dan talks for a few minutes with his section boss. It seems

to be one of the rare nights when nothing is going wrong. So many times the accidents seem inexplicable, like cables getting cut or drills snapping in half or gears getting fouled up for lack of oil. All foremen suspect that these problems are intentional, but they rarely accuse the men of causing them.

A foreman knows that if he accused a section crew of causing a breakdown, he might drive the men into stronger anger. Most foremen prefer to announce the breakdown to the front office and make repairs as soon as possible. The front office, still under pressure to run more coal, puts pressure back on the foreman, who solves the problem by rushing the equipment back without proper maintenance or by neglecting certain safeguards to health or safety. Dan sometimes feels reminiscent about the old days, when the bosses could fire the workers on the spot—but when the company also knew how to keep the mine in working shape. All that expertise was lost a long time ago, Dan thinks.

The men take a break for supper and Dan joins them. Some of his favorite people are on this section: Calvin; gentle little Jimmy Reid; Deacon Lambert, who preaches on Sunday; big Lew Begley, part Cherokee, with two fingers missing on his guitar-picking hand; two younger lads just hired on recently; and Frank Baker, who works only about half the time now, since the doctor warned him to quit completely with his black lung. Frank looks weak and tired these days.

The men dip into their lunches, glad for the silence from the machines, not eager to talk.

From the outward tracks of the mine shaft (the "out-by") comes the clackety-clack of another jeep, its light bouncing off the curving wall. Then the men spot the powder car, with its cylindrical body, filled with crushed rock.

Two men operate the powder car—slippery Dean and young Mark Tierney, with his long hair and stories about all the college girls he has topped over in Ashley County. They make an odd couple, the sly little fellow who has never strayed far from Sorrow Creek and the younger man just back from Vietnam. But they get along well together.

The powder car looms close now, its rounded body caked in rock dust. Seen in the flickering light, the two supply men look squinty-eyed, like raccoons seen at the side of a dark road, like gondoliers gliding past on a dark Venetian night.

Tierney nods his head.

Dean makes a quick waving motion with his right hand.

Then—splat! Water splashes off the back wall, showering all eight men seated below. And the powder car glides out of sight.

"He's done it again, doggone him," Jimmy pipes up.

"That Dean, I'm gonna have to kill him," Lew Begley booms in his deep voice.

Dan Sizemore starts to laugh, his body shaking helplessly, his voice going "yah-yah-yah" above the miners' curses. They all know that Dean has pulled another of his beautiful tricks, filling up a plastic sandwich bag with swamp water, where the men piss and the frogs play, and flicking the bag into the midst of his poor dumb buddies.

"I mean it, Dan, I'm gonna kill the little bastard," Begley roars.

"Now just be thankful that Dean is back to his own wonderful self," Dan says, when he recovers his voice.

The men still muttering behind him, Dan gets in the jeep and heads toward other sections. But nothing is going wrong tonight. He patrols the shaft until 10 P.M., when he begins the forty-minute haul to the outside. Soon he is in the chilly outside air. He looks up and sees the sky is clear tonight.

Back in the office, Dan makes his positive report to the expediter. Great night. Dan has almost a free hour to wait for the shift to end. He glances up at the safety poster and sees

Dean's portrait again. Can't be letting Dean get too cocky, he thinks.

Dan grabs a bucket of engine grease from a supply shed. He walks into the deserted bathhouse. He spots Dean's basket by the immaculate cowboy boots he wears. He lowers the basket until it touches the floor. Then Dan applies the luscious-looking red jelly, deep inside the toes of both boots. Then he raises the basket to the ceiling again and replaces the bucket of grease.

At midnight the man-trips clatter out of the tunnel. The men slip off the carts even before they come to a stop. Greasy and muddy, their eyes glittering, they scramble wordlessly toward the bathhouse, past the men of the hoot-owl shift now assembling.

In the bathhouse they tear off their grimy clothes and rush into the shower, thickening big-muscled men mixed with scrawny old-timers with sagging gray flesh and young men with firm pink bodies. In the shower they pour huge quantities of Joy liquid soap (the only soap that cuts the grease) on each others' heads, like boys in a swimming pond, or they turn the water too hot or too cold on an unsuspecting buddy. Then, muttering threats through the foggy air, they rush off to get into clean clothes.

Fastidious little Estill Dean puts on his spotless green overalls and his spotless green work shirt, which he wears in his off-hours. Then his socks. And then his boots. And then he starts walking toward the door. But his footsteps become tentative, like a man slipping in mud, and he puzzles over the sensation inside his boots. Slowly Dean sits down on the bench and removes one boot. The red lubricant stretches from the toe of his sock to the inside of his boot, like a strand of red molasses. He pulls off his other boot. Same thing.

"Who did this?" Dean asks, thin-voiced. "Come on, who did this?"

The men come rushing from all corners of the bathhouse

to see what has happened to Dean. They roar with laughter when they see.

"Who did it? I swear, I'll get him," Dean says, looking around him, his sallow face turning red.

Dan has refrained from laughing. Foremen, after all, are supposed to be dignified. He ambles over toward Dean and says, "Dean, I can help you find the culprit."

Dean looks upon Dan Sizemore as a voice of friendly authority.

"How?" he asks.

"Go get me a yellow pad and pencil," Dan says.

"How is that gonna help?" Dean asks.

"Well, looky here, Dean, all you've got to do is tell me the name of every man you've ever thrown a water bag at. I'll write down the names and I'll bet you a bottle of the finest moonshine your family ever smuggled down off Sorrow Creek that the culprit is somewhere on that list."

The men see Dean's mind working hard. They see him mentally filling out one side of the legal pad, the names spilling onto a second sheet, on and on.

"Aw, shit, Dan Sizemore, you ain't no goddam help at all," Dean whines. Then he puts on the boots and more or less glides out into the night.

"Wee-yaw, wee-yaw," Dan roars, along with the other miners.

Slowly, still bellowing, the men depart. Nobody asks who doctored Dean's boots.

4

One A.M. Still in darkness, Dan Sizemore has traded his miner's lamp for a flashlight.

Walking up the dirt path in silence toward his house, he holds the flashlight in front of him, looking out for mudholes or pony shit or night-crawling snakes.

He walks slowly, his heart pounding. He thinks of the pictures they showed him over in Doc Rasmussen's laboratory in Beckley, the rotted lungs with the coal particles permanently implanted, making the heart work overtime. They said he had second-stage black lung, but he decided not to apply just yet, because foremen make more money than people on black lung.

The path goes past the Bailey house. David still thinks it was the Baileys who shot their dog two summers ago. Past the old Montgomery house, the memories of Lester's fat, rotting corpse. Past Ralph Thomas's house, the quiet people with their sudden fights.

Up ahead the Sizemore house looms below massive Milan Ridge. The house is illuminated by the streetlight they put in that vicious summer, when the pack of thugs was lurking in the bushes, looking for trouble. The last bugs of the season dance in the glow of the streetlight.

The lights are on downstairs, a good sign. Dan likes company when he comes home. He climbs the stairs to the screened-in porch, past the swinging couch where the family has spent so many pleasant times together. In the living room, off to the right, Bobby and Vicky are still awake. The room is full of cigarette smoke. The radio is blasting rock music. Bobby is reading the Tri-Cities paper, his long hair hanging into his eyes.

Vicky is wearing white karate pajamas, sitting on the floor, bending forward, rocking to the rhythm of the radio.

"Evening," Dan says.

He takes the pistol out of his pocket and places it back on the mantelpiece. He puts his lunch pail in the kitchen, makes a cup of coffee from the constant coffeepot, walks back into the living room.

"Where's your mama?"

"She went to bed early," Bobby replies. His voice is soft. "She said she was tired from writing a sociology paper."

Dan sips his coffee. They all smoke. Vicky shifts to her back, throws her legs backwards, past her head, yoga position. Nobody speaks. Dan gets up and turns off the radio.

"Whee-yew, that seems mighty loud after that quiet walk through the grove," he says.

They are silent again. Then Vicky finishes her exercise, squats Indian style, leans forward to talk. Her hair is yellow like her father's and her facial bones are rugged, like his.

"We got a letter from Pete and Chris today," she says,

drawing out her syllables. Her voice is the most Southern accent in the family. "Said it near snowed last week."

Nobody has seen the two oldest Sizemore boys since Dan's vacation last July. They are planning to drive up again at Christmastime.

"They've got tickets to a hockey game," Bobby says.

"Remember we saw them play Montreal last Christmas?" Dan says. "I thought Pete and Chris were going to fight some of those Montreal fans."

Vicky goes into the kitchen and fixes a bowl of cottage cheese and chives. They all dip crackers into the bowl, crumbs and flecks of cottage cheese falling on the cigarette-scarred table. Nobody talks until the snack is gone.

"How is school?" Dan asks, addressing either child. Bobby does not respond. He has not shown any interest in school since the principal suspended him two years ago for his long hair. Dan and Margaret went to court with a Civil Liberties Union lawyer and got some district judge to get Bobby reinstated. But Bobby just about gave up on school after that. He lives for his summer camping trips and his visits to his brothers in Toronto.

"They're trying to get us to pay for our diploma," Vicky says. "I told them I wasn't going to pay fifteen dollars for something I had earned."

Dan smiles. True to the Sizemore tradition, Vicky despises the dreary little people who run the Bradshaw County High. Only gentle Mary-Ann has ever held back her contempt for the pitiful undereducated teachers, who have never been much farther then Tri-Cities in their lives.

The Sizemores have all been marked in school anyway, ever since Dan and Margaret organized their citizens' group back in the late 1960s, trying to keep the free lunch program from being used by the politicians. After that the Sizemore children heard their parents called "Commies" in classroom discussions.

Sometimes the school episodes are ludicrous. One of Vicky's teachers was lecturing on *The Scarlet Letter* and was just getting to the point of what the letter *A* stood for when a school board member came running in from the hallway shouting, "You can't teach that! You can't teach that." The teacher had to call for the principal to explain to the board member why a twelfth grade could stand a brief discussion of adultery.

Most of the children have heard about sex by the twelfth grade.

Just a few weeks ago a couple was discovered using a vacant classroom for something other than homework. The boy remained in school but the girl dropped out when rumor got around that she had paid the boy for his services. Dan noted that if the incident had taken place in a classic novel—as sometimes such incidents do—it could not have been safely taught within the confines of Bradshaw County High.

Vicky also tells about one of her teachers back in the eighth grade, who used to quote articles in the *National Enquirer* as a source for his information. Although she has many boyfriends, who stop over on weekends and take her for drives, Vicky says she has no immediate interest in getting married and starting a family like her girl friends. Some days Vicky talks about being a lawyer. Other days she talks about working in the mine. Margaret says she is a dreamer. She is her father's daughter, for sure.

Dan puts out his cigarette and glances at the clock.

"Almost two o'clock," he says. "Reckon you-all gonna feel like going to school in the morning?"

"I guess so," Vicky says, half-drawling, half-yawning.

Dan feels tired now, the tension from the mine going away. Wordlessly, the three of them drift upstairs, take turns in the overworked single bathroom, scatter to the four bedrooms that hold the eight Sizemores still at home.

Dan undresses silently, so as not to disturb Margaret, who is sleeping deeply in her jeans and polo shirt. The bright moon outside casts its light into the bedroom, glistening off the rifle barrel.

Dan slips into bed. The air is absolutely silent, inside and outside. The house sleeps.

5

Dan enjoys these evenings with his older children, enjoys any time with his family, because his memories of his own father are shrouded in darkness. He can remember his father leaving for work before dawn and coming home after dark, his clothes and body blackened from the mines, for there were no bathhouses in those days.

He is a true son of the coalfields. His earliest memories are of piles of burning slag along the narrow roadways, strings of wooden coal camps along the bottomlands, men rushing back and forth from work.

When Dan was born his family was living in Pike County, Kentucky. But his father soon followed a job across the Tug Fork into Mingo County, West Virginia.

There has always been plenty of commerce across the Tug Fork. A Mingo family named Hatfield used to ford the Tug

Fork just to vote in Kentucky elections. A Pike County family named McCoy never much cared for that practice. The slaughter had just about stopped when the Sizemores settled in Mingo County, but the county was still preoccupied with murder.

Dan remembers his father taking him downtown into Matewan and showing him bullet marks on the train depot, where the Baldwin-Felts agents had been killed in a shootout trying to break up a strike in 1920.

The Baldwin-Felts agents were hated by union people for their roles in earlier slaughters in Cabin Creek and Paint Creek, West Virginia, and also at Ludlow, Colorado. When the Stone Mountain Coal Company brought them in to stop the rising United Mine Workers, on May 19, 1920, Sheriff Sid Hatfield tried to get them to leave. Nine men were killed in the resultant shootout and Sid Hatfield would later be ambushed when attending a trial in a neighboring county.

"You'd hear all kinds of stories," Dan recalls. "My father told me about one Baldwin-Felts detective who was trying to climb a stairs to a dentist's office but the dentist hit him with a jar of dental material and killed him. I don't know if that was true."

It was hard getting the true story in the coalfields, then or now, because the newspapers and the school systems were often controlled by the coal companies. In Dan's schools, it was safe to discuss the Civil War by the 1920s, but the teachers never referred to the more recent strikes or shootouts or the adventure over in West Virginia, where General Billy Mitchell practiced for building an air force by flying low over the striking tent colonies. Dan would get most of these stories from his father.

Although he was a coal boss himself, Ed Sizemore always talked sympathetically about the union to his young son. Relatives have suggested that Ed Sizemore once tried to organize a mine back in his native Tennessee but was run out of the state. Once he started bossing in Eastern Kentucky, he never got involved with the union again.

There were a lot of things Dan learned belatedly about his father. One time a friend told how Ed had been a Mississippi River gambler. This was a total surprise to the Sizemores, since Ed never turned a card in his own house. Yet Dan sensed that his father had lived with danger early in his life and was desperate to live out his life quietly. Mining was dangerous enough without all the extra frolics.

"My daddy didn't talk much," Dan recalls. "I used to think he was cold-feeling. But now I see. The man was worn out, broken down from working. There wasn't much he could say. But I didn't understand in those days."

Dan's father had started working in the coal mines when he was eleven years old. The coal companies liked to hire young boys because they were so quick and agile. They could straddle a coal chute and separate the good coal from the worthless shale. They could hunker down in the tunnels and operate the trapdoors that separated sections of the mines. And they could scramble on the outside of the coal cars and operate the hand brakes, where no fully grown man with slowed-down reflexes and an acquired sense of danger would want to work.

With good trapper boys and good brake boys, the company could run twenty or thirty coal cars at a time, without stopping on most grades. But as the companies pushed for production, numbers of boys barely into their teens would be crushed between the cars and the walls of the mine, an occurrence that was rarely treated as a supreme tragedy by the grizzled older miners.

"The older miners hated the kids," Dan says. "They called them smart alecks. You know how those older miners must have been—set, stodgy, working at a snail's pace. Along comes some smart, aggressive young kid doing the toughest work in the mine, twelve or fourteen years old. After a while, the kids would get promoted to motorman."

Ed Sizemore survived his early careers as trapper boy, union organizer, and Mississippi gambler. Then he became a private contractor, hired by the Cold Creek Coal Corporation, owned by the paternal Jos. McGuire Company.

It was still a time of the pick and shovel. Miners were paid for the coal they produced, not for the hour they spent trudging through the mine shaft to the working face or the hour spent getting out. Nor were they paid for separating the coal from the waste. If they wanted to make their two or three dollars a day, they had to load their sixteen tons and hope the weight man was a friend of theirs and was giving an honest count that day. If there was an accident or a delay and they couldn't load their coal that day, it was too bad.

"There was a standing rule: 'Bring the cut of coal—or bring your tools,'" Dan recalls. "That meant, if you didn't clean up the coal, you could haul your tools out of the mine because you were finished."

As boss of eight or ten men, Ed Sizemore would make as much as twenty seven to thirty dollars a day, enabling the Sizemores to live in a nice home near the Cold Creek No. 6 Camp near Matewan. The miners lived in the traditional main camp at Cold Creek: identical wooden buildings with the inevitable wooden company store within easy reach.

Coal camps have long since been discredited by their image of exploitation: the scrip that was redeemable only at the company store. Social awareness has reduced life in a coal camp to something close to indentured slavery. Some of it was. But many other mining families had never had it so good before moving to a coal camp and they were sure the good times of the 1920s were going to last forever. Even Dan Sizemore, turned bitter by events in his adult years, remembers his childhood as good years.

"My memories of the coal camps are mostly of summertime," he says. "Swimming, ball games, company paternalism. My dad and a couple of older brothers were working, we had two or three cars, including a new Nash, that was the real thing in those days. My mother was ill much of the time, but she knew the need for children getting out in the fresh air. My father worked six days a week but we always went out riding or picnicking on Sunday.

"The country was still pretty except for a few slag heaps. There was no strip mining then. They'd dump the slate on the mountains, with no thought to the future. The subsidence from the mines would turn the big rivers the color of beautiful orange pop. But the country was still prettier than it is today. We could swim in the creeks in those days. Hell, you can't swim anywhere in the mountains anymore. Strip mining has ruined it all."

For many mining families, whether the older British settlers or the newer immigrants from Italy or Eastern Europe, life in the coal camps gave them a sense of identity they never had before.

The camp at Cold Creek had a community hall where the company showed movies several nights a week. The old projectors would grind away with Tom Mix, Buck Jones, Douglas Fairbanks, Sr., and the Gish Sisters, the operator changing reels manually. Dan and his buddies would sneak into the theater in the middle of a crowd, saving their dime for soda and popcorn. When they sneaked into the movies on Sunday, Dan's religious Methodist mother would whip him for it. But that didn't stop him.

Weekends were sometimes enlivened when Dan's father drank too much of the local moonshine. Whiskey seemed to tap a special rage within him and other miners on their brief weekend of Saturday night and Sunday. When Ed Sizemore started a commotion, the townspeople would stay out of his way. A town official once put him in a casket and bolted it, just to keep Ed out of trouble for a few hours.

"That didn't disturb his Irish soul one bit," Dan recalls. "Being part Irish, I think my daddy always expected to wake up and find himself in a casket. But most of the time he just worked."

While Dan's father was off running coal, the children were raised by two strong women—his mother and his father's stepmother. To this day Dan talks with awe about rugged old Grandma Roxy, gallivanting around the countryside with animals and children following in her wake. He marvels at the strong tobacco she chewed and the bitter pipe she smoked. One time Dan and his brother Frank smoked Grandma Roxy's pipe while sitting in the branches of a tree. The heady aroma literally knocked them out of the tree, at the feet of Grandma Roxy, who whipped them worse than any playground enemy ever could.

"It was more than just love or admiration we felt for Grandma Roxy," Dan recalls. "It was more like fear."

Dan's mother was descended from a Pennsylvania Dutch lumbering family that had moved down the Blue Ridge when she was a girl. Jeannette Sizemore had only a third-grade education, yet she influenced the family heavily with her firm morality, manners, and knowledge.

Some of her worldliness may have been gained from her friendship with a wealthy businessman named E. B. Baker, for whom she had worked.

"He actually used to drive into our camp and bring his wife to visit my mother when she was ill," Dan recalls, his cynical tone fading at the memory. "That really meant something to us. It was like my mother wasn't some little slavy who worked for him. It was like he was visiting kinfolk."

The visit was no small thing in the mountains, where comfortable townspeople rarely have much interest in families from the coal camps or from the head-of-the-hollow. There is so little affluence to go around that any small success is quickly resented. And the wealthy usually grow contemptuous of the poor; there are few liberal trappings, no radical-chic cocktail parties for militant hillbillies, in Appalachia. Usually the lines are drawn with all the charm and subtlety of medieval master and serf. That's what made the visits from E. B. Baker and his wife so special. Dan isn't sure why the Bakers socialized with his mother; he likes to think they were rare, decent people who recognized a kindred soul in his disciplined, ambitious mother.

Sitting in the living room, listening to the worldly Mr. and Mrs. Baker talking with his mother, Dan got his first feelings for the outside world. It was more than he got from his teachers, ill-qualified and poorly paid, who taught by rote and were not inclined to deal with inquisitive children. The school library had a few battered Zane Grey Westerns, some outdated textbooks, and little else.

It was an environment that traditionally has destroyed young minds in Appalachia. Dan's recent experiences have convinced him that this is no accident. He believes that the coal companies and their politicians contrive low school taxes to keep the education inferior.

"Do you think the coal companies want us to be educated?" he snorts. "What happens to their labor force if the poor mountain man ever figures out the score? This is all part of a plan."

Yet Dan managed to escape that cycle, largely because of his mother and the huge supply of books owned by his older brother, Ford, the oldest of the seven Sizemores.

"I never got to know Ford very well," Dan recalls. "I wish I did. But he was always out early, working in the mines, from the time I was young. When I got older, he already had lung trouble—miner's asthma, we called it in those days, the same old shit, black lung. He moved to Greenbrier, up on the plateau, where it's supposed to be better for your lungs. He left his books behind."

Ford had all kinds of books, French short stories, Russian novels, written by men whose names Dan cannot pronounce perfectly to this day, since he read their work long before he ever heard their names spoken. Dan roared through Charles Dickens. He loved travel and history books.

"I'd see pictures of the Statue of Liberty from the same place the immigrants saw it, from the harbor. I wanted to visit there. I'd see pictures of the Welsh miners. I felt a great kinship for them, with their odd caps, their flat Appalachian faces. I'd feel a great thrill thinking about our kinship."

Already coal had a hold on the young man. He could see his father and brothers going into the mines; he heard of the deaths and the injuries. Yet even though the men complained about specifics, they had a feeling of purpose in digging out the coal, day after day. The radio told them that America was having a roaring good time in the 1920s. And when the coal-camp band played a concert on Saturday night, it wasn't hard to convince themselves that they were part of the national party.

"It wasn't all bad in those days. Sure, the men worked long hours. That wasn't good for them, was it? But they never thought about social conditions back then. Those Saturday night square dances, when all of us would be out having a good time using the company facilities, nobody ever thought, where the hell are the black miners? What were they doing? It never worried me in those days. That's what worries me now."

Sitting in his living room today, surrounded by books about DuBois and photographs of Martin Luther King, Dan recalls his first awareness of black Appalachians.

The blacks have been mining coal just about as long as whites, some as slaves, some as strikebreakers, always the last to be hired and the first to be fired. They lived in separate colored hollows, sometimes even worked in separate sections of the mine (called "Africa").

But the coalfields are not the Deep South, either. Because most of the mountain farmers had no need for slaves, there was little identification with the Confederacy. The mountains became a vital backbone for the underground railroad; most counties sided with the Union. And when the Civil War ended, blacks and whites worked out their uneasy compromises.

"We used to have fights with black kids, not because we hated them but because it seemed like the thing to do, to get approval," Dan recalls. "One time a principal of a black school came over and talked to me. He said his kids had enough trouble getting to school without buses. He made so much sense, I don't think I ever gave the kids any trouble after that.

"Another time, we stole the tar paper off the roof of a guy named Jeff Hayes. Usually you could do that with a black man because the whites wouldn't take his word if he complained. But my daddy wasn't like that. My daddy beat us good. He gave us what we needed.

"My mother was a wonderful woman, but she was full of little prejudices. Not only toward blacks. Hell, she was that way about Catholics. She'd say, 'My, so-and-so is a nice person—for a Catholic.' My mother felt the colored were different somehow.

"I remember this black fellow, Powell, he was the same age as me. We were good friends but we went to different schools. One time I heard that King Solomon was black but I refused to accept it. I told Powell that King Solomon must have been a different kind of black—certainly not black like Powell. I'm sure I thought I knew what I was talking about, at the time."

It was a time of national certainty. But that all came to a crashing stop one afternoon in 1929, when Dan was twelve years old. The Depression hit the coal industry as quick as any. The miners didn't have a chance to hear about layoffs—they just happened. At Cold Creek the McGuires saw there was no sense in evicting anybody and the benevolent owners tried to see that a family wouldn't starve to death. But then the companies began to feel the pinch and they began turning off the electricity.

"If my daddy didn't have a disability from the armed services, I don't think we could have gotten through," Dan recalls. "One family I know of lived in a cave. You'd eat bread. You'd hoard corn. The cars started rusting away because you couldn't get parts. You'd walk everywhere. Walk, walk, walk. You'd walk to the grist mill and pay them a few pennies for grinding the precious corn."

With few of them in danger of being evicted, the miners waited out the long term of Herbert Hoover, growing their traditional gardens on the steep hillsides, comforted by the feeling of a shared crisis. In later years, Dan would be torn inside by being laid off by himself. But in the early 1930s everybody was laid off.

"You'd think the Depression was a horrible time," Dan says. "But it wasn't, in a way. It was the best-natured fellowship you ever saw. It was a lark, a picnic. Life went on as usual. You'd have your political and religious arguments, like always.

"You'd play baseball with an old pick handle, an old baseball, taped over and over. You'd have maybe one catcher's mitt. Lots of guys would go without gloves. But you'd always have ball games to take your minds off your misery. We played and played."

The few freight trains that still moved all carried a cargo of hoboes, dozens of them. Dan and his friends would throw apples at the hoboes, who would catch them and eat them.

Then in 1932 the country elected a man who would become a savior in the hearts of most mountaineers.

"Franklin Delano Roosevelt had a kingly manner," Dan says. "God, how they loved him. Those guys would have killed you if you made a snotty remark about the god of the coal miners."

When the WPA was founded, the men of Cold Creek began working again, two or three days a week on dilapidated trucks, getting a ticket worth six dollars at the company store. The mines would work one or two days a month. But there was a feeling of movement. Then the mines began to run full shifts and Dan's father went back to bossing for the last few years of his life. And because jobs were still not plentiful, Dan stuck it out and graduated from high school in 1936.

"By that time they were paying four dollars for a straight eight-hour day," he recalls. "The days of the free cleanup were over. You worked your eight hours and went home. Buddy, that was the life. As soon as I could, I got me a job in the mines. I really thought I had arrived. Buddy, I was some punkin in those days."

6

When Dan went to work in the coal mines in 1936, he vowed he would be a tough coal boss like his daddy. He was going to rise in the company and move more coal than any boss had ever moved.

They put him outside, working on the motors, until they discovered the qualities that Dan Sizemore wanted them to discover. In Dan's words, "They recognized hardness, callousness. I was the exact type of bastard they were looking for."

So they sent Dan inside the mine with a clipboard and a stopwatch, doing time studies on the section crews. The men were no longer paid individually by the coal they loaded; that system was long since discarded. But that didn't mean that a hard-eyed coal boss couldn't drive them toward higher production.

Dan would single out one of the better miners who would

set the standard for the section. If the leader could shovel thirty huge scoops per minute, Dan would expect every other miner to shovel twenty-five. That way, Dan would get between twelve and fifteen tons of coal per man per shift. The men would shovel furiously, with their foreman shouting at them and young Dan Sizemore studiously timing their efforts.

There really wasn't much choice for the miners. Either they met the standards or they were unemployed in a one-industry region. There wasn't much difference between Cold Creek, West Virginia, and some old Roman slave galley—except that stopwatches hadn't been invented in the Romans' day.

The miners had nobody to turn to. Although the United Mine Workers had enrolled most of West Virginia following the union wars of the early twentieth century, the union had little impact on the daily boss-miner relationship. If the miner tried to talk reasonably with his section boss, he was usually fired. If he tried to reach a higher boss at the company office, he usually found the executives unwilling to listen, since they were under pressure from the home office to produce coal. Even at the paternal McGuire mines, the miner was at the mercy of one man—the foreman. And Dan Sizemore felt proud to be the best foreman in Cold Creek No. 6 Camp.

In later years, Dan Sizemore would cite his younger days as an example of the "master-slave relationship" he says is so natural to the southern mountains. He would recall hideous examples of the miners' servility—one man knocking on his door on Christmas Eve to whiningly present him with a Christmas ham. But at the time Dan thought it was his due.

Dan fired a lot of miners but the one he remembers best is Collins.

Collins was a local boy, five years older than Dan, who was then a punk of twenty. Collins had been a football hero at

Matewan High; Dan had often gone to watch this rangy halfback with curly reddish hair romp over Pikeville or Williamson teams. Collins's family had always owned stores in the coal valley, but they had fallen upon bad times during the Depression and now their men were forced to work underground.

This was in the thirty-six-inch seam of coal in the western fringe of the state. The miners worked on their hands and knees, the roof rubbing against their backs, because it was much too expensive to excavate the mine to a six-foot height.

The miners wore kneepads like basketball players. They would go home and soak their aching joints in hot water, cursing the day they settled in a valley with three-foot coal. And the next day, knees creaking, they crawled back into the mine, like children playing under the porch.

But Collins had a special agony: he had two knees that had been distended and torn while running with the football for Matewan High, listening to the cheers of Dan Sizemore and his little buddies. Collins worked a month in Dan's mine, never complaining when the foreman warned him he was falling behind. Then Dan timed him several days in a row and found he was the low man on the shift. Without a word to Collins—yet directly in front of him—Dan told the foreman that Collins had to go.

Never saying a word, Collins put down his tools and crawled toward the mouth of the mine. To this day, Dan doesn't know whether Collins crawled all the way out the four miles or whether some foreman took pity on him and offered him a ride in the man-trip. But that was none of Dan's concern. He had done his job.

Dan never saw Collins again. He heard that he went to work in one of the nonunion "doghole" mines that paid half wages for twice the danger. Most miners insist they would rather rob a bank or go on welfare than work in a doghole. But desperation drives them there.

Soon after Collins was fired, Dan's superintendent told

him, "Dan, you're the only one I can trust to fire people." That made Dan feel wonderful, back in 1938.

The region was starting to boom, giving thanks to Franklin Delano Roosevelt and keeping one eye on the impending war in Europe. Dan's father died, but Dan had no trouble supporting his mother and his widowed sister, Edith, with enough left over for good times on weekends. When World War II broke out, Dan and his brother Frank were the only two Sizemore boys eligible for the draft. The law said one of them could stay home and support the women and work in the coal mines while the other would have to defend the country.

"Frank had this burning desire to join the navy and see the world," Dan recalls. "I had different desires. Even when Frank got a commission, there was no jealousy on my part. I was too busy mining coal and dating two girls in Cold Creek at one time. With all the men away at war, the girls never objected. And if you missed a shift because you had a particularly nice girl you hated to leave, the company would say, 'You shouldn't do that.' But what could they do? They needed me."

Dan's free-living ways ended shortly after the war ended. While trying to hide from an agent of a foreman's union, who was trying to enlist Dan and his buddies, Dan drove over to Williamson, with its movie theaters and restaurants and dancing joints. There he met a pretty red-headed cashier from the Roberts Coal Company store. Her name was Margaret Hamilton and, except for being ten years younger than Dan, they had pretty similar coal-camp backgrounds. They met in October and were married in December and rented their own foreman's-size house for fifteen dollars per week. For a while, Dan's sister was pretty upset about losing the support from

Dan. But the couple settled down to enjoy a postwar boom that reminded people of the 1920s.

Freed from its war-making chores, American technology went back to the drawing board and came up with, among other things, a Joy Loader that moved alongside a seam of cut coal, scooped it up, and dumped it on a conveyor at its rear. Dan took delight in realizing that he had outlived the age of the pick and shovel.

But for each step forward there was a step backward. Old Joseph McGuire, who had run the company with paternal benevolence for as long as Dan could remember, had now acquired enough money to retire and enjoy his older years. To prepare for that time, he brought in his nephew, right out of Yale University, to help supervise the foremen.

Young James McGuire smoked his cigarettes with a limp wrist, like a woman, and talked in a squeaky voice. One of his first discoveries was that Dan Sizemore was known as the toughest foreman in the mine. One day James McGuire, right in front of a foreman meeting, informed Dan that he had the blackest labor record of anybody in management and would have to mend his ways.

Dan stood up and told James McGuire, "You don't know the half of it. I've straightened out that No. 6 mine for your uncle. I packed a pistol and backed down one big son of a bitch who was causing us problems. And now you tell me I've got the blackest labor record?"

Dan further informed the young nephew that he was leading the production lists, which meant that most of the men worked eagerly for him.

After that outburst, Dan figured he was done at the McGuire mine, so he began applying for a job as inspector for the U.S. Bureau of Mines. But old Joseph McGuire came around one day with his wife to bid farewell to Dan and Margaret and announce he was retiring to Florida. Old Joseph McGuire even said he would put in a good word for Dan with the nephew. A few days later James McGuire came around, cigarette dangling from his limp hand, and said he meant to keep Dan working. One month later he transferred Dan to the worst section in the mine.

Dan took it for a while. He even pushed his section into the top of the production charts. But James McGuire's hired hands would follow Dan around, looking for mistakes. One Friday Dan told James McGuire that he was quitting. The mine superintendent was shocked but the young boss said, "Let him quit."

That afternoon Dan was sitting around his front porch, drinking a can of beer, thinking of the ten years he had spent running coal for the McGuires, working three or four straight shifts without pay if they had a problem, keeping all their secrets to himself, firing employees who couldn't produce, and all the time leading their production charts.

While Dan was brooding, the McGuire mine had a ventilation crisis, a sudden crack in the underground mine wall, unleashing a pocket of explosive methane gas. The best foremen in the mine were needed immediately to help ventilate that section, before one spark blew the mine all the way across the Tug Fork. One of the bosses came by and said, "Oh, there you are, we need your help over at the mine."

Dan looked at him coldly and said, "I reckon I'll be moving out of this camp by the end of the week." And he opened another beer for himself. The McGuire mine did not blow up. Dan had a job with another company in three days.

The Franklin Coal Company, now a subsidiary of a giant oil company, is one of the leading producers of coal in the United States. When Dan joined it, Franklin was already owned by outside interests who hired "a cold-eyed bunch of devils" to supervise their investment.

Franklin was heartless in some ways but they knew how to give incentive to their foremen. All three shifts working the same section would share any incentive bonus. That would guarantee that one shift would perform all necessary maintenance and leave the section clean for the next shift. The opposite is true at Big Ridge, which is also owned by an outside corporation but pits one shift against another shift as if they were rival baseball teams.

Soon after joining Franklin, Dan set a record that still stands. With ten men working on one section, he moved 237 cars in one eight-hour shift, worth approximately 1,200 tons of coal. The cash rewards enabled Dan and Margaret to rent a comfortable house, large enough for their family, which seemed to grow once a year.

But there was something missing at Franklin. In the old days under Joseph McGuire, Dan had felt like a human being. The top executives at McGuire had been his fishing buddies. But at Franklin most of the top men were eager corporation types from outside the region, trying to claw their way out of the mine office and into the plush home office up north.

As an old-fashioned mountain coal miner, Dan never felt at ease in the Franklin camp. The easygoing atmosphere of Cold Creek No. 6 had been replaced by the high-pressure tactics of a suburban subdivision, where career is more important than friends.

Instead of visiting casually from one porch to another, day after day, the Franklin crowd seemed to stick to themselves except for an occasional wild party on Saturday nights, when the men would drink hard and the wives would perform stripteases on a dining-room table, as everybody seemed desperate to work away the competitive tensions of the week.

Dan was hardly a prude, not after his long years as a bachelor during World War II. But the antics in the Franklin camp were alien to him and Margaret and they started to withdraw from the social life.

The Wooton No. 26 Camp was thoughtfully laid out in vacant land between two giant coal-cleaning plants. When the company was cleaning its most valuable resource, the dust swirled around the valley, like a snowstorm on a negative film. The Sizemores didn't dare hang their clothes outside but bought an indoor washer and dryer instead. The camp swimming pool had a thick scum on it every morning. Margaret would sweep off the front porch every hour so the kids wouldn't track dust into the house. They kept all the doors and windows shut, but the dust would filter in anyway, through the keyhole, under the doors, everywhere. From a distance, it seemed that they had a lovely grassy yard. But when Pete and Chris, the two oldest boys, romped on the lawn, their clothes would be black from the soot.

The Sizemores were living better than they had ever lived before. When Dan was promoted from foreman to superintendent, he lost his outlet to incentive money until he bargained his way back to foreman. They always had a new car and bundles of groceries, paid for in good green cash. Nobody was paid in scrip anymore and nobody owed his soul to the company store. Whatever they spent was their own free choice.

But for all the money, the Franklin foremen stayed desperate. The company hired one roving foreman for every shift, with no section for him to boss, no incentive bonus to be made. His job was to lurk around the mine, hoping to discover somebody else in error, hoping to gain his own section to boss. Sometimes Dan thought about Collins and all the poor lads he had terrorized with his stopwatch, back in the late thirties.

Another way Franklin kept its men insecure was by transferring them around. After Dan set the production record at Wooton No. 26, Franklin moved him over to No. 25. That was in 1958, when the economy seemed to be slipping during the second term of Dwight D. Eisenhower.

Dan didn't worry about moving at first. But he was never fully accepted by the bosses over at 25. They had their own little groups and they resented men from 26.

One day Dan was called into the front office and handed a

slip of paper. There is nothing subtle about the coal industry. The slip of paper was a beautiful pink color and it performed the duty that pink slips have always done—it informed him in a few words that his services were no longer needed.

"All the times that hard little Dan had used these words against other men," Dan recalls. "Now they were being used against him. I didn't say a word. No statement was necessary. But it was the hardest jolt in my life. I was hurt and surprised."

The next day Dan went to the company bank to draw out some savings for the next few weeks. The bank clerk seemed surprised and asked why Dan was taking out his savings. When he heard the news, he showed Dan a cutoff list that did not have Dan's name on it.

What had happened was that a clique from No. 25 had done some last-minute politicking and protected their buddies by getting Dan added to the list.

Dan thought about fighting it. But he wasn't worried. He knew he could move coal better than any bastard that ever lived. He was confident that Dan Sizemore could get a job, anyplace, anytime.

7

When the man handed Dan the pink slip, Margaret was already pregnant for the seventh time. The boys were so sure she would deliver another boy that they even picked a name for him—George Crowe, after the star first baseman of the Cincinnati Reds, the "home team" for most mountaineers.

The Sizemores joked about the layoff for the first few days, confident that Dan could find a job anywhere. But when he made his first casual moves toward other jobs, he found there were none. It was 1958. The coal industry had crashed during "Ike's Recession," as the miners called it. The rest of the country felt the tremors of the recession, but the miners were caught in the rubble. There were no national crash programs like the ones FDR had started during the Great Depression. It was just the poor mountaineers, with their politicians unable to get the kind of relief the region needed.

Mines closed everywhere, while the United Mine Workers obligingly took the pressure off many marginal operations by cutting entire areas out of union membership. In Eastern Kentucky the union abandoned almost all its membership, leaving the miners at the mercy of doghole operators. The overall registration in the UMW, which had been 600,000 in 1947, plummeted down toward 200,000.

In other areas the companies consolidated their operations, lopping off whole sections but continuing to move the coal. Dan could still see his neighbors drive off to work at Franklin every morning.

He decided that was worse than everybody being out of work during the Depression. Misery loves company; Dan and Margaret didn't want to starve by themselves.

Money started getting scarce. They had just traded in a good car for a big new station wagon. The Christmas bills were hanging over their heads. They had some savings, so they could remain in their house, but each day Dan would drive around to the personnel offices and then come home and tell Margaret and the kids "no jobs."

Then came the day he stood in line to collect his first unemployment check. There must have been four hundred men standing on Third Avenue in Williamson, men who would rather have been risking their lives under the ground than standing on line, taking the dole. When Dan came home with the welfare check, he rushed to the bathroom and threw up.

He likened his feelings to *The Grapes of Wrath,* which he had read several times. He remembered the pitiful Okies, how they used to apologize to the policemen for being picked up on the highway, how low they felt. That was how Dan felt, when he stood on the unemployment line with all his poor fellow miners.

Dan developed a feeling that maybe the layoff was his fault. Maybe if he had been working harder, they would have kept him. He knew it was foolish to feel guilty, remembering the production records he had set, remembering the hard-driving bastard he had willingly been. But the guilty feelings persisted.

He did a lot of pacing, a lot of walking. He stopped reading. All he did was chase rumors, from county to county, not taking any chance on missing a job. But the trip home was worse each day. He began to think the neighbors were talking about him because they were working and he was not.

Unused to scrimping after years on a foreman's salary, they began to economize. Margaret spent eighteen or twenty dollars per week on food for the eight of them. She would stock up on mackerel at nineteen cents a can, mix it with corn meal, and fry it into little round cakes.

They drank powdered milk and ate long red beans—not the tasty kidney beans that mountain people love doused in gravy, but something harder. Margaret would soak the beans and cook them for hours, but they would stay so hard they bounced on the plates. Cabbage was the big crop that year: the farmers gave it to them free; they ate raw cabbage and cooked cabbage.

Margaret's mother gave them coffee and cigarettes, which they could not afford. They would smoke Pall Malls, not because they liked them but because they were so long. They could smoke them, put them out, and smoke them again later. Dan's sister sent them chunks of tasteless government cheese, so much of it that they'd feed it to some puppies living under the floorboards of the house. The cheese would stick to their teeth and the family could hear them growling, trying to get their jaws unstuck.

Poverty brought home lessons of shame to Dan and Margaret. Until the layoff, they had lived for themselves, often neglecting their relatives. Other people were more generous to Dan and Margaret's parents than they had been. Perhaps because Dan and Margaret felt they never had much when they were younger, they felt they didn't want to share it with others. But when Dan was out of work, they saw how thoughtful other people were. They began to understand how fast your dignity and self-respect could disappear. They had formerly complained about people on welfare who were too lazy to look for jobs. But now they discovered how it felt.

Sometimes Dan would spend their last five dollars on a tankful of gas, just chasing down a rumor. He began looking outside West Virginia, crossing over into Kentucky to look at the dreaded doghole mines.

One time he heard about a mine opening in Leslie County, Kentucky, one of the poorest counties in the United States, with a high birth rate and a corresponding death rate. Leslie County once went through three sheriffs in one week during the early union wars of the 1920s. With much of the coal land owned by the Ford Motor Company, the county was still violently antiunion, when Dan decided to try his luck there.

But some of Leslie County's roving citizens spotted his West Virginia license plates as he crossed the line from Perry County. When he pulled into a gas station, three cars surrounded his and several gruff men leaped out of each car, demanding to know what an outsider was doing in Leslie County. Undoubtedly they must have thought he was a union organizer. Dan rapidly decided he did not want to work in Leslie County. He told the men, "Fellers, as soon as I get my gas tank filled, there will be no more reason for you to ask me why I am here—because I am going to drive right on out again."

Dan never returned to Eastern Kentucky. Twelve years later, he would hear the name of Leslie County again, when the doghole Finley mine alongside Hurricane Creek blew up because of illegal blasting operations, killing all thirty-eight men inside.

Dan stayed in West Virginia for a while. Like most mountain people, he did not want to leave the hills to find work. Yet so many people were forced to leave Appalachia that some counties, like Leslie County, would lose up to 25 percent of their population between 1950 and 1960, as the people drove north to the factory towns like Dayton and Cincinnati and "Dee-troit City."

Dan had never been north in his life. Cities were still a

far-off dream to him. He still thought of himself as a mining man, ready to work wherever there were minerals to be dug. When he heard about a hard-rock uranium mine opening up in Grants, New Mexico, he took some of his dwindling savings and drove all the way to New Mexico by himself.

Grants, New Mexico, was out on the desert—hotter and dryer than Dan had ever imagined. Yet on the first level of the underground mine it was so damp that men wore rain hats and raincoats. The boss offered Dan a job in a few months, but he could not afford to wait around. Many West Virginians waited around for a uranium job or moved over to potash mines. But Dan needed money real fast. He drove straight home again, the whole trip another horror.

Back home, Margaret was nearing the end of her pregnancy. She had remained unusually slim, gaining only seven pounds during the nine months, too busy feeding her children and keeping the house running to think about eating right.

The money got tighter. One time they ran out completely and couldn't pay the water bill of $2.50. The company sent a man around with orders to shut off the valve right away. It was a Monday; Margaret said she could pay the bill by Friday, but the man said he had orders and he was real sorry. He took out a shovel and pick and found the valve all covered with dirt and turned off the water.

When the water stopped running, Margaret felt herself grow numb. With six children, how was she going to manage without water? For the first time since the layoff, she felt she had reached the end.

Just then the man next door came over and said, "Mrs. Sizemore, I don't want to seem like I'm trying to embarrass you, but I couldn't help but see that fellow turning off your water. If you'd like to, you'd be welcome to attach a hose to my outdoor faucet and run it to your house."

Margaret knew how hard it was for that man to make the offer, because most country people she knew felt timid about butting into other folks' affairs. But she gratefully accepted his offer until she could pay the water bill on Friday. She wanted to have water running for when the baby arrived.

A desperate man will often do something he shouldn't.

After six months without work, Dan took a job with a doghole mine in Rocky Creek down in southern Logan County, not too far from Buffalo Creek. (In 1972 a dam constructed by another concern, a Pittston Coal Company subsidiary, would disintegrate, flooding Buffalo Creek hollow and killing at least 120 persons.) The people who ran the mine at Rocky Creek were not known for their safe practices. They had a loose-rock formation above the seam of coal and the men worked in constant danger of roof falls.

Dan put in two weeks as foreman on the day shift. One day he kept the section idle because the roof-bolter was slow in securing a newly opened area. When the superintendent told Dan to move the cutting machine into the new area, Dan refused. He said he wasn't going to move men or machines into the area until the roof was secure. The superintendent got mad and ordered the roof-bolter out of the section and ordered the cutting machine moved in. He was under orders to move coal that day; no foreman was going to slow him down. The super told Dan he would have another crew erect timber supports while the cutting machine was loosening coal.

"Be my guest," Dan snapped, moving outside the mine. "I don't want to have any part of it."

A few hours later, while working in the office, Dan heard an awful scream. He rushed inside and saw a machine helper lying on the floor, his face bleeding badly from a falling rock. The wound stopped bleeding after a while and the injury was not serious. But Dan knew it could have been worse.

Rather than roof-bolt the section, the superintendent called for some jacks, to move the fallen rock, and some laborers to carry the injured miner outside, so work could continue. When Dan asked the superintendent what had happened, the man admitted he had not timbered the section first.

Dan turned 180 degrees and walked out of the mine. He found the mine foreman inside the office.

"Let me tell you two things," Dan blurted. "First is, I quit. Then let me add, I don't give a good goddamn what that super tells you, that little S.O.B. got that boy hurt. Remember now, I've quit, so it don't make any difference to me. You can cut coal any way you see fit. That bastard is going to kill somebody. But it ain't going to be me."

Dan drove straight home and told Margaret he had quit. She started to cry. She said she knew he did the right thing. But she cried and cried.

By quitting, at least Dan would be home for the birth of their seventh child in June. They did not have enough money for the delivery bill, which must be paid in advance in many Appalachian hospitals. However, Margaret's mother gave them the money and they reserved a space for delivery and one night in the hospital.

The boys had been right when they predicted another boy: he had bright red sideburns and blue eyes. Dan and Margaret chickened out about naming him George Crowe; they named him Kevin, instead. After one night in the hospital, they wrapped him in the new green recovery blanket they had purchased for Green Stamps and they took him home.

That first night home, Margaret's mother stayed at the house while Margaret tried to sleep. She didn't dare let Kevin sleep in the same bed with her: she was so tired she was afraid she'd fall asleep and roll over on the baby.

During the night, Margaret's mother heard Kevin making

some noise and she tried giving him a bottle. But the milk just trickled out of his mouth. She woke up Margaret when the baby seemed to be a little blue in the face. They decided Kevin was chilled and they covered him up. In the morning, he was making faces and they thought he was having gas pains, the way babies do. But around 11 A.M. his arm turned a deep purple. Dan rushed them over to the hospital.

The doctor put Kevin in the incubator and watched for the signs of the arm turning blue. For a while there were no danger signs. But then Kevin went into convulsions and the doctor said his lungs were putting extra pressure on his heart. Dan and Margaret took the news with a silent grief. Kevin died at 7:15 P.M.

Dan and Margaret left the hospital that evening, bitterly asking themselves if the doctors would have given them better care if they could have afforded it. They buried their son on a hillside near their home and wondered what was next.

Later that summer a traveling salesman Dan barely knew mentioned a new mine opening in Southwest Virginia. Dan learned that an old boss of his had a good job down there. He looked at the map and located Bradshaw County, just about as deep in the mountains as any spot in Appalachia. It was a good four hours from the familiar camps of West Virginia, but Dan got in his car and drove down there the next morning.

At first it seemed like the same old routine, a bunch of hapless miners from Eastern Kentucky and West Virginia standing around, begging for work. The Big Ridge Coal Company had its office in a beautiful old log cabin. Dan managed to gain entrance to the office, where he found his old boss, Jacob Davis. The man seemed glad to spot Dan but said he could not hire him for several months, since they were only in the process of leveling the land outside the mine. They would have to work with construction crews, opening the mine to the first crosscut, before they could afford to hire miners.

Dan told Davis it had to be now or never. Dan insisted he could work on a construction crew as well as any mine crew. Perhaps sensing the desperate tone of Dan's voice, Davis smiled and said, "Dan, I'm gonna stretch a point. Take your papers and go down to Milan and get a physical. If you pass that, I'll get you into the United Construction Workers and you'll have a job."

In the makeshift hospital at Milan, Dan passed the physical. He notified Davis and headed home to tell his family the good news. He was so happy, he stopped at a grocery store on the first ridge he crossed and bought himself a six-pack of beer. Drinking cold beer on the four-hour drive, he arrived home laughing and crying for joy.

The new job meant the family would have to move eventually. But for the first few months Dan would live in a hotel, while looking for a place to house the family. His job was to supervise the burning of all the brush and trees the crew had leveled from the mountainsides where Big Ridge would put its mine. He worked from 3 P.M. to daybreak, seven days a week when they let him. He needed as much overtime as he could make.

While the fires were raging, Dan would see local people arriving in trucks, begging for the timber. He knew he was burning enough wood to construct a dozen homes, but the company had ordered him not to let anybody carry away a single stick. The people would beg for the lumber and Dan would say, "Buddy, I do whatever the company says."

He knew if he let them in, they would start robbing coal from the exposed highwalls of coal, where the dynamite men had shot the face of the mountain away. The local people would sneak around the roadblocks when nobody was looking. They carried small sticks of dynamite that could loosen a truckful of coal with one quick blast. Or they'd carry a small auger, which they strapped to their bodies with an elaborate leather chest-harness. They'd come in pickup trucks, wives and kids and grandparents and shovels, and they'd load half a ton of coal in a matter of minutes.

Dan's job was to keep them out. He could see the company's point. Hell, he'd spent his life protecting one company or another. But now for the first time in his life, he could see the people's point, too. They were starving up on some scruffy ridge somewhere and they could see the company burning a mountainside of timber. Dan could understand the misery that drove these people; he had just taken a nine-month taste of it himself.

But Dan owed money to everybody back home. It was getting worse now that he had a job and people were troubling Margaret to repay them. She tried to spread the payments around and, after a while, they saw they would all get paid. But Dan was not going to endanger his job by letting the people cart away firewood or coal.

After a few months, Dan rented a house in a farm section of Bradshaw County. Then he hired a truck and moved his family down from Mingo County. He felt a great sense of relief when his family was reunited in Virginia.

Early in 1959 Big Ridge started running coal and Dan was a foreman again. But he was a different kind of foreman now. He had done a lot of thinking in the past year. He had evolved his "master-slave" theory about the typical Southern working attitude. In the past, if a miner had come to him with a problem, Dan had always figured it was the worker's fault rather than the boss's. But the layoff had shaken him loose of that blind loyalty. He saw how the capitalist coal companies "rewarded" their employees by laying them off when the dividends fell. He decided that from now on, his loyalty would be to himself and the men who worked for him. Never again would he be the "raw son of a bitch" he used to be.

Tuesday

8

Four hours later, after Dan went to sleep—long before daylight has filtered over the ridge—the kitchen is occupied again.

Still in the dungarees and polo shirt she fell asleep in last night, Margaret Sizemore pours the day's first cup of coffee. As if coming out of a dose of ether, she rests her elbows on the kitchen table, her face impassive below her dark-red hair.

No matter how tired she feels, Margaret gets up by six o'clock every morning because it is the only time she can be alone. All the rest of the day will be spent traveling to college, studying, sitting in class, caring for her family. This is her hour, to smoke, to drink coffee, to think.

The kitchen is cluttered from the snacks last night—coffee cups and ashtrays and random plates scattered around the sink, a few cold potatoes that somebody left uncovered in a pan.

The kitchen has been Margaret's home base for twenty-five years, from the time she gave up her job in the Roberts camp store and married Dan Sizemore, the cocky, yellow-haired foreman from Cold Creek. Then it was children, eight of them plus Kevin, who just barely got home before he turned sick and died. All the meals prepared—and the children seen off to school—and then Pete and Chris going, one by one, over the border. And now her two little grandsons living down in Morgantown, whom she hardly gets to see.

A grandmother and a college student. Last year she and Dan decided she should study nursing so she would have a profession in case he got hurt or the mine closed down.

She was terrified of college at first. After twenty-five years away from school, she was afraid she wouldn't be able to read or write. She was afraid of being humiliated in front of all the generation of bright young college students she had heard about. But soon after she started, she saw the reality of a little community college in the back hollows of the coal country.

Most of the children were products of the region, sons and daughters of nonunion miners or operators. These children talked about the unions as if they were an instrument of the devil, as if Tony Boyle stood for all unions, with his misuse of union funds, with his goon squads. She had tried to tell them about Dan's friends in the reform movement, but the children didn't want to know, most of them.

And Vietnam. Even after all those years of Johnson and Nixon, the children still supported the bombing in Southeast Asia, still opposed giving amnesty to the boys who didn't "do their duty."

Margaret used to come home from school shaken beyond tears, numbed by the hardness of these young people. Where was the peace generation she heard about on the radio? They sang the songs, they wore their hair long, but their hearts were hard, most of them.

Sometimes in debates, when she was feeling wise and mature, Margaret tried to stimulate their thinking with the Socratic approach of questioning. But when the subject got around to amnesty, she felt herself growing emotional, thinking about her two sons, so far away, the vicious hatreds that would keep them from ever coming home to the mountains again. And she would drop the discussion. It was too much, too raw.

Then there were the policemen. They wore their uniforms, with pistols tucked into their holsters, taking college courses as part of their training, sitting in class, intimidating people with their crew-cut air of authority. Although they kept to themselves pretty much, Margaret often wondered what they might do if they did not agree with somebody in a debate. Or what if the handful of angry blacks on campus staged some sort of little demonstration? Or what if somebody got up and defended the boys who went to Canada? Who knows whether these policemen would act like college students or like cops?

Dan suggested Margaret carry their rifle to class one day and explain that was part of her working uniform as a hillbilly mama. But Margaret didn't think they would see the humor over in Morgantown.

What might have happened yesterday, if the policemen had sat in her sociology class, when three white boys and three black boys debated each other in open hatred?

The white boys shouted, "Why did you come over here? We didn't want you in this country?"

And the black boys shouted, "We didn't want to come here. When we get control, we're gonna give it to you twice as bad."

The few moderates in the class had tried to form a bond between both sides. But it had been impossible. She was not surprised by the white boys; she had heard talk like that all her life. But the open hatred by the young blacks . . . when did that start? Blacks had always seemed like the most mature and patient people she knew.

It was a black man who had helped Margaret join the human race, that evil year when danger seemed to follow the Sizemore family. When beautiful Charlie York came to stay at their house and the thugs started making remarks. When Margaret learned how to use the rifle, when she vowed to risk her life to protect the black friend. Now to hear these boys in her class talking of Armageddon . . . it made her suffer so bad.

Margaret's decision to fight for Charlie York had shaken her in ways she hadn't imagined. She had always supported Dan's theories in a feminine, well-meaning, Dan-knows-best kind of way. But after the incident with Charlie, after all their battles, Margaret had changed from passive to active. And when they talked of her learning to be a nurse, she signed up for college, terrified of the work, but dedicated to her goal.

After making supper, after clearing the dining-room table of Monopoly games, old coffee cups, boys' trucks, and Dan's writings and papers, with the boys wrestling and shouting, with the rock music blasting through the house, Margaret and Mary-Ann studied together, night after night.

In the first semester, they both made the dean's list. After one year, Margaret was invited to join the college honor society. And she was accepted into the exclusive nursing program. There are times when she thinks longingly of the social science courses, the humanities courses, the classes about ideas and action that she will never get to take. But nursing may get them closer to their two sons.

At 6:30 Margaret hears Mary-Ann stirring on the stairs. Mary-Ann is blond, like Vicky, but her features are softer, more shy. Around the house, Mary-Ann usually wears dungarees and a sweat shirt, looking rumpled and comfortable. But this morning she is wearing a pastel skirt and blouse, her hair neatly combed, looking as graceful as college girls always look in glossy magazines.

Wordlessly Mary-Ann fixes coffee and a piece of toast. How nice she looks, Margaret thinks. The two of them seem to match, dark red and light blond, two equal halves that communicate without saying a word.

Margaret decides to wear her tweedy maroon skirt and jacket today, to dress up. She debates whether to wear stockings. If there is any mud at all on the path, it will spatter on her stockings, making her look like something just off the ridge for the first time. The way she picks up mud, she can always tell where she's been sitting in school from the tracks of hardened mud that caked off the bottom of her ripple-soled shoes.

At the end of a dirt path, Margaret loves the solitude of their rented home, loves the peace of the front porch, with no traffic whizzing past, like down in the coal camp. But she misses the neighbors, the kind of neighbors her family had when she was young, people who would visit from porch to porch, drop in for coffee, watch the babies if somebody got sick, talk about the pie auction down at the Baptist church.

That was back when she was Margaret Hamilton, from Logan County, West Virginia, who believed in Jesus Christ and Franklin Delano Roosevelt and felt she would always be part of the coal camp. But ever since Dan's layoff in 1958, all their moves led them toward this lonely dirt path, turning them away from the church they now despised, turning them away from the bomber presidents they now despised, turned them against the leaders of Bradshaw County—and turned many people against them.

Margaret thinks about an invitation she received recently to a Tupperware party down in the camp. She thinks she might like to attend, if she can finish her term paper by then.

It would be good to let the women in the camp remember that Margaret Sizemore is not some kind of ogre.

Margaret clears her throat. She says her first words of the day.

"Ruby's having some women over for a Tupperware party Friday night," she says. Her voice is soft and tentative. "Would you like to go with me?"

"Maybe," Mary-Ann says. Her voice is barely audible.

They sip their coffee a little longer. Then Margaret slips upstairs and changes into her maroon outfit. She takes a chance on the mud and she puts on her sheer stockings, feeling lavishly dressed for a change. Then she wakes up Vicky in one room, Bobby in another room, Edward and Gene in a third room. When they begin to stir, Margaret decides there is a fighting chance they will all rouse themselves out to school in another hour. She cannot wait around to help them, as she used to. At five minutes after seven, Margaret and Mary-Ann start walking down the dirt path, stepping around a few soft, muddy spots.

9

Tuesday morning is cloudy, the sky hanging low over the mountains. Dan wakes up slowly in the kitchen, taking his first cup of instant coffee. The house is filled with the heavy thumping of rock music, coming from the stereo set in David's room upstairs.

Shortly there is a thumping in the hallway and Dan looks up to see the tallest member of the Sizemore family. David's hair is cut short, almost a crew cut, in direct contrast to the dangling locks of the other Sizemores, both here and in Toronto.

David is smiling eagerly.

"Morning, Daddy," he says.

"David," Dan says slowly, pronouncing the name with loving inflections, about four syllables worth, as if the name were a title like "Your Lordship." David comes over to him, hunching over slightly from all the years of trying to stoop down to the lower heights of his younger classmates. Dan puts his arm around the waist of this muscular nineteen-year-old man in dungarees and a polo shirt, with a touch of beard starting to show on his face. The two men purr with happiness.

"Daddy, why you smoke?" David says, sniffing the ever-present cigarette burning in the ashtray.

"I've told you, David. So's I won't feel out of place when I start breathing inside the mine."

"Daddy, don't smoke no more. Gives you cancer, right, Daddy?"

"Damn right it does, David. Tuberculosis, too. But smoking never gave me black lung."

"Don't smoke no more, Daddy," David says. "When I get my Volkswagen Camper, I'm not going to let anybody smoke. Not Vicky, not Mary-Ann, not Bobby. Just my mommy."

"That's right, David."

Dan stares up at his son. David surely knows how to tease. He absolutely drives Vicky up the wall, picking on her for the illogic of being a vegetarian who smokes two packs a day.

Now why would a boy who can tease like that, who can grasp complex personal relationships, not be able to retain his lessons in school, grow weepy when left alone in the house? The times he was tested, the doctors said it was mild damage, affecting certain parts of the brain. But how much? Dan and Margaret wish they had been able to take David to specialists in one of the big cities. But it had always seemed too difficult to make the appointment and take the grueling trip out of Southwest Virginia, where there is no facility for testing or training retarded children.

Or maybe David's got the right idea. Dan sometimes thinks that the gentle boy took one quick look around him at this rotten world, with its capitalists ripping off their huge profits at the expense of the workers and the politicians screwing the people who had elected them, and crawled back into his shell again, like the baby chicken crawling back into the egg.

Or maybe it was the long night when David was trying to be born, when he couldn't turn around inside, and the country doctor gave Margaret the shots to make her sleep—or maybe the shots were so the doctor himself could get a few hours' sleep.

In the morning Margaret spent two and a half hours on the delivery table, convinced she and the baby were both going to die. She told Dan later, "I'd wake up and they'd be struggling and I'd feel myself rising above it all, looking down at them, wondering what was happening. I could see the nurses were concerned. Then I went to sleep again."

When Margaret awoke again, the nurses didn't show David to her for a while. When they brought him in, he had four finger marks on his forehead and two big forceps marks on the side of his head. The finger marks went away eventually, but the forceps marks are still there, under the short haircut.

By the time he was four, David wasn't talking and he wasn't trained. The local doctors couldn't explain much. He started school at six, like the other kids, but showed no aptitude for schoolwork. The doctors said he was not educable, so the Sizemores began to keep him home.

But there were special things about David.

When the Sizemores first moved down from West Virginia—David was around five—they rented an old house that was supposed to be haunted. Both Dan and Margaret had grown up bearing the old mountain legends of "haints" in old houses, ghosts in the mine shafts, but they felt they were much too intelligent to believe in spirits anymore. So they closed off parts of the house they couldn't use and they went about their business.

One room they kept locked was an old storeroom right off

the porch. David used to put his eye to the keyhole and laugh so happily to himself that the Sizemores felt glad just hearing him. One time Dan asked David why he laughed.

"Lady dances in there," David said.

The Sizemores smiled patiently at David and paid it no mind. But a few weeks later, while cleaning out the attic, they discovered an old scrapbook that had belonged to the previous owners.

When David handled the scrapbook, he found a picture of a pretty young woman. He became excited and indicated she was the woman who danced in the storeroom downstairs.

The Sizemores made some inquiries and found that the woman in the picture had died when she was nineteen years old and her spirit was supposed to have remained in the house. They next opened the storeroom and discovered a sewing form of a woman. But David insisted he had seen a whole woman, dancing around the room. After a while, neither Dan nor Margaret felt prepared to disbelieve their son.

David stayed out of school four years when the Sizemores moved to their current home in Milan (pronounced "MY-len" by the residents). Since no facilities existed for retarded children, state law said he could attend the county grade school.

By this time David was a teen-ager, towering above his classmates. Many of the boys were from broken homes, or overcrowded homes, or homes where brutality was never far from the surface. They soon sensed that the gentle giant in their classroom was quite different from them. David did not enjoy playing ball or fighting or aggressing the teacher. All he wanted was attention and some chores to perform. The boys began teasing him, poking him, testing him to see what made him different. One time David came home with a bleeding wound deep in his thigh, because a younger boy had jabbed him with a pencil to see if he would cry. Another time a group of boys stole a watch that David dearly loved. Each time Margaret would go down to school and complain to the teachers. But the teachers shrugged and indicated they could not control the activity of the rough-and-tumble boys every moment.

The issue was complicated because Dan and Margaret had become politically active in the county, forming a citizens' committee to fight for a free lunch program. The school board and the county judge had viewed the Sizemores as dangerous opponents in a region where dissent is equated with open subversion. So the Sizemores found a not very receptive atmosphere when they tried to discuss David's plight.

As he reached his middle teens, David began to defend himself, discovering that one shove could bowl over one of his smaller tormentors. The teachers began to complain that David was actually the bully in several fights. Yet despite the fights, David loved school and showed a warm affection for his teachers, whether they returned it or not. And he looked forward to graduating from the eighth grade, with the familiar procession and celebration, just before his twentieth birthday.

But in the fall of 1972, when David reported to start the eighth grade, the officials sent him home. They said he had reached the age of nineteen and, under state law, was no longer eligible to attend grade school. Dan came home from the mine one night and found Margaret weeping in the living room.

"They found they could hurt us through David," she said.

So David stays home now, listening to rock records all day long, fretting nervously until Dan wakes up, waiting for the house to fill up with his brothers and sisters each afternoon. Dan has encouraged him to get the paper from the mailbox, to run errands to the local grocery store, and to learn how to cook.

"Mr. Simpson told me 'Go home, David, go home.' Why, Daddy?"

"Aw, David, don't worry. You had fun in school. But now

you're a big man. You don't need school anymore. You stay home with me."

"I'm going to work in the coal mine like you," David says.

"That's right, David. You're gonna ride one of them jeeps with me one of those days. I'm gonna take you over there maybe this Saturday and you'll ride into the mine with me. But right now, why don't you make me some of your delicious scrambled eggs?"

They walked into the kitchen. Dan sits at the table and reads the paper. Over by the stove, the man with the strong arms chuckles to himself as he breaks an egg and drops it into the frying pan.

10

Dan has been talking about the number of miners who suffer from black lung or pneumoconiosis—the irreversible presence of coal dust in the lungs of a miner. Today he has planned a visit with two brothers, Jimmy and Isaac Reid, who both have lung problems. They live over in Sara-May, one of the many coal camps or mines named after women.

The ride will take thirty minutes going north on the main road. The car sways back and forth as we maneuver the sharp turns parallel to the twisting creek. Most mountain drivers learn early that there is no use trying to make time on the winding roads. Dan rolls along around thirty miles an hour and talks about the brothers.

"Jimmy is a little fellow who works on my shift, lovely little guy. Isaac doesn't look anything like Jimmy. Isaac is a huge feller. A legend around these parts. He used to mine coal six days a week, then he'd pitch a doubleheader for the camp team on Sundays. Isaac would pitch the first game of a doubleheader with his left hand and then come back in the second game and pitch with his right hand. They'd make trips as far as Pittsburgh or Birmingham to play coal teams or steel teams. They'd bet money on Isaac and he didn't let them down very often."

When Dan talks about baseball, he gets enthusiastic the way city men talk about their vacations in Europe or their football tickets. In the modern urban areas, it is fashionable for men to pretend they understand professional football and to call baseball a game of the past. Maybe it is. But it is still very real as Dan Sizemore negotiates the twisting valley road.

"Lord, the baseball that used to be played around here," Dan says. "People forget what baseball meant in the thirties. It was a Depression sport. It was just the right sport for people sitting around and watching something for three hours. In the big cities, men would watch ball games for fifty cents, kill a whole day that way. And the company teams would build a kind of spirit, hold the men together during the week.

"We had a lot of guys like Isaac Reid—great big studs who could have made their mark in professional sports. Some guys would make it out. But there'd be other guys who would prefer the security of the lunch pail and the six-day week. People would always wonder if they could have made it."

Dan's eyes grow brighter as he recalls the great baseball players from the coal camps—Paul Derringer and Bobby Bowman from over in West Virginia, where he grew up. Doc Edwards from Pigeon Creek, near where Dan lived with Franklin. Selva Lewis Burdette from Nitro, West Virginia. The Niekro brothers from Ohio. And Tracy Stallard, who lives just a few ridges away from Dan.

I remember Tracy Stallard. He was a big handsome pitcher who served up Roger Maris's sixty-first home run in 1961 and later pitched for the New York Mets. Sportswriters used to call Tracy a "Virginian," conjuring up an image of a courtly Thomas Jefferson or George Washington in spiked shoes. Yet I never associated Tracy with the coalfields of Virginia until one time when I was driving near Dan's house and saw a whole cemetery full of Stallard markers. I inquired and found Tracy's home, a pleasant house on the ridge, his mother rocking and knitting, his daddy working over at the mines. He seemed stunned to see me so far from New York. There was a painting of him in his Met uniform, taking up an entire wall in the living room. Tracy talked hesitantly about his baseball career. By that time he was pitching in a minor league in Mexico during the summer. He didn't say it, but I could understand that pitching in the minor leagues of Mexico beat working in the coal mines, like his father. That was a few years ago. I hope Tracy has managed to stay out of the mines.

Dan continues talking about the old coal camps: "The best ballplayers were often these great beautiful black men who were not allowed to participate in the white man's leagues. My God, some of the athletes we saw."

Dan raves about a man named Bennie Moore, a giant of around six foot four and 230 pounds, with a deep beautiful voice. He died young from a kidney infection. Dan remembers watching him play against the touring colored teams like the Homestead Greys with Josh Gibson and Satchel Paige.

One time the colored teams were playing a preliminary game over in Williamson, where the Saint Louis Cardinals had a minor-league team. Branch Rickey had built up his "farm system" with hundreds of outstanding young players—men like Terry Moore, Stan Musial, and Del Rice. The miners would watch them play one summer in Williamson, read about them in Rochester or Columbus the following summer, and then see their names in a Saint Louis box score the summer after that.

On this particular summer evening, Bennie Moore and his colored team was playing the Cuban Giants. Bennie hit a homer over the deep center-field fence, clear into the Tug Fork. The next time he came up, all the Cardinals scampered out of their clubhouse to watch the giant black man swing. Sure enough, as all the white miners cheered like crazy in the grandstands, Bennie Moore slugged another home run, right into the Tug Fork. Dan says that Stan Musial himself never reached the river.

"We had some great ballplayers in the coal camps," Dan concludes. "But it's safe to say that Isaac Reid was one of the greatest."

Dan drives into Sara-May, a pleasant, flat town nestled at the confluence of two river valleys. He finds the brightly painted brick house where Jimmy Reid lives. Jimmy is out front, repairing a wrought-iron fence on his stairway. In his high-pitched voice, Jimmy invites us into the spotless house, its kitchen appliances more modern and numerous than the Sizemores'.

In the kitchen, a portrait of John Fitzgerald Kennedy is hanging above the refrigerator. A caption underneath says: "Ask not what your country can do for you. Ask what you can do for your country."

Jimmy fixes a cup of coffee and says Isaac will be over in a few minutes.

"Isaac is feeling better. He had trouble breathing for a while but I think he's feeling better now. He's still trying to get his black lung benefits. I think if they'd give him that, he'd feel a lot happier."

Black lung. The nickname is only a few years old; but then again, so is the recognition of the ailment. Only recently have doctors and government officials conceded that coal dust interferes with breathing and puts extra strain on the heart. The first federal compensation law in this country was enacted in 1969—thirty-five years after Great Britain's.

The door rattles and Isaac Reid walks in. He is obviously an athlete, with a broad neck and thick arms and a thin white sport shirt stretched across his massive chest. His hair is jet black, but he eases himself into a chair like an old man. He is not yet sixty years old.

"Yeah, I'm better now," Isaac says, the words coming out slowly, without force. "The doctors cut me open to take out a cyst. They said that might help."

Isaac unbuttons his flimsy shirt, displaying his huge and hairy chest, marred by three livid surgical scars.

"They told me they could see the black lung in there. But I still can't get my compensation."

He buttons his shirt again while we all take a hasty gulp of coffee. Isaac's broad flat face and his weak voice remind me of another wasted athlete I saw when I was a little boy—the ruined body and voice of Babe Ruth, wrapped in a camel's-hair coat, rasping his farewell at Yankee Stadium a few months before dying in 1948. Isaac looks something like Babe Ruth.

"I know when I ruined my lungs," Isaac continues. "I was just a kid when I started, fifteen years old. We were taught that the dust didn't bother you. Your foreman told you that. Your doctor told you that. You just cough it out of your system, they said.

"I was working over that way"—he gestures toward the opposite ridge—"and we ran into four hundred feet of sandstone, forty-eight inches high, twelve feet wide. Me and five other guys operated jackhammers, day after day, in that white sandstone. The dust was so thick you couldn't see your body. At the end of the day, you had to take a pencil to dig the white stuff out of your mouth.

"Of the six guys on the job, I'm the only one who lived this old. One guy died at age thirty-five; he just smothered to death. I had to pull him out of the mine many a time. They all died of natural causes. 'Cause I lived longer than them, a few years ago I got paid off for silicosis. But they didn't pay me for black lung.

"I put in forty continuous years in the mines without missing a paycheck. I worked overtime and holidays and extra shifts. The way I figure it, I worked sixty years out of my forty. I kept going until the doctor told me flat out, 'We're gonna have you in here with a heart attack if you don't stop working.'

"But I can't get the Social Security people to pay me for black lung. I can't get them and the doctors to agree on anything."

Isaac's story is common. All over the coal region, weakened middle-aged men shuffle from hospital to Social Security office, from doctor to lawyer, in an attempt to get certified. Men carry tattered envelopes with all their medical records, opening them with damaged hands, eager to tell their legal struggles to any sympathetic stranger. Widows carry their husbands' records in their pocketbooks, stopping in offices for one more try at retroactive payment.

The compensation is recent but the complaints about chest pains, about shortness of breath, about early disability and premature death have been present for generations.

"They used to call it 'quimsy,' a hog illness," Jimmy says. "They'd tell us it was just shortness of breath when were wheezing like a hog. In England they call it 'grinder's rot.' I've performed artificial respiration on older guys with the skin stretched across their face.

"They told us to chew tobacco," Jimmy continues. "That keeps the stuff from going down. Heck, if you're out of the mine for a week of vacation, you might get your nose clean. But I can always pull up a bunch of black stuff and spit it out. It's disgusting what's inside of you."

Dan has been glowering quietly in his chair, letting the other men do the talking. Now he has something to say.

"I remember a rotten little state senator saying you could get coal dust from dirty sheets but not from working in the mine. But shit. The coal company owned him like they own everybody else."

The men firmly believe they have been sold out by almost everybody. They might not have expected the coal companies to be concerned with the dangers of black lung, since that would have cut into their profits. But the men don't recall help from other agencies. They insist that the United Mine Workers never warned them about black lung until long after other individuals had raised the issue.

When Tony Boyle was running for reelection in 1972, he insisted that the UMW had been trying to alert the miners for over twenty years, trying to present papers on the subject to medical and governmental authorities.

"But they wouldn't listen to us," Boyle frequently whined. When asked for evidence of his efforts, Boyle fell silent. That was only one of the reasons why he lost the election.

In fact, the increased awareness of black lung sprang out of southern West Virginia in the early 1960s, often spurred by "outside" doctors who had no emotional or financial ties to the coal industry.

One such "outsider" was Dr. Werner Laqueur, a suave European who had arrived in Beckley, West Virginia, in 1949. Dr. Laqueur is not interested in politics. He once told me his story quite hesitantly, not wishing to sound like an activist.

Dr. Laqueur is a pathologist. He does autopsies. When he first arrived in Beckley, he had numerous occasions to discover what lay in the lungs of coal miners who had died of heart stoppages or other "natural" causes.

"I started to do autopsies," Dr. Laqueur once said. "Clearly, we did not have silicosis, which is a nodular disease, quite different from pneumoconiosis. In 1952 we stopped using the term 'silicosis' and began calling it pneumoconiosis. This conflicted with the standards of the West Virginia Board of Compensation."

This means that the miners could not be paid if disabled by pneumoconiosis.

"I started to collect lungs. In 1960 I said that 83 percent of miners had pneumoconiosis, in any form, light or bad. I said 29 percent of my miners had died directly from pneumoconiosis. . . . Lawyers tried to introduce this pneumoconiosis [to the compensation board].

"How could I prove it? Coal burns at 425 degrees Fahrenheit. You burn the lungs and prove there is coal dust present."

But burning the lungs of the dead miners did not immediately help the living. Coal dust still swirled uncontrolled within the mines and men began to wheeze and slow down after ten or fifteen years underground.

"I asked, when does pneumoconiosis disable a man? I cannot say whether this person was disabled. But it is hard to believe this does not cause functional impairment," Dr. Laqueur continued. "Yet fifteen years ago, doctors called it an 'anxiety complex' in miners. . . .

"You can say, well, the man can still breathe. But how much? And you don't say what happens to the air [in the lungs]. The Public Health Service was hesitant to say we have a problem."

When doctors like this presented their findings, other officials were not quick to get involved. Dr. Laqueur was no crusader. Perhaps his European background in the years leading up to World War II had taught him to keep out of the limelight; I don't know; I never asked him. But he seemed like an honest doctor, a man of facts and service. And other doctors came along to make a crusade of it.

One of the crusaders was Dr. Donald Rasmussen, who arrived in Beckley in 1962, a pulmonary specialist from out West, with no background in coal problems. Working in the same hospital as Dr. Laqueur, Dr. Rasmussen soon recognized the social implications of the autopsies and the broken, living miners. Armed with Dr. Laqueur's studies and his own pulmonary knowledge, Dr. Rasmussen began screaming.

Soon miners began to listen to this stocky young doctor with the bright red beard and the intense shining eyes of a Van Gogh self-portrait. They began to flock to the pulmonary clinic in Beckley, abandoning their coal-camp doctors, convinced they would get a full, fair diagnosis from the new man.

One of the miners who visited Beckley was a shy chap named Arnold Miller, from the bloody Cabin Creek section. Arnold Miller's handsome face had been repaired by plastic surgery after the landing at Normandy. Then he had gone into another dangerous business—coal mining. By the early 1960s, as he turned forty years old, Arnold Miller discovered he couldn't climb hills anymore without wheezing. When he visited Beckley, Dr. Rasmussen told him why.

Arnold Miller, who had silently endured machine gun bullets for his country, decided he had taken enough. He and his friends organized the Black Lung Association, and they made their affliction one of the hottest issues of the decade in the coalfields.

A few other doctors joined the cause, including Hawey (Sonny) Wells of Princeton and I. E. Buff of Charleston.

Dr. Buff, well into his sixties, began driving around the coalfields, carrying plaster models of lungs, pictures and diagrams of rotted lungs and overgrown hearts. And for the *pièce de résistance* he would bring out a slide of a real blackened lung, showing the coal dust embedded deep in the tissue of a miner, long since dead of "natural causes."

Dr. Buff was a missionary, a close cousin to the country preachers who predict hell and damnation for sinners on the radio and in the little churches. He roared at the miners that "some of you ain't gonna live until my next visit." He told dark tales of a grand scheme by the government and other doctors to keep miners ignorant.

Dr. Buff's exhortations were scorned by even some of the doctors who agreed with him on black lung. But he helped stir up many miners who had never questioned their breathing conditions before.

Not surprisingly, the Rasmussens and Buffs were soon attacked by the established medical authorities. Some doctors said that miners' biggest problem was emphyzema, caused by too much smoking.

"There is no epidemic of devastating, killing and disabling man-made plague among coal workers," said Dr. Rowland Burns in a message to the Cabell County Medical Society in 1969. Dr. Burns also referred to "false prophets and deluded men who present their hypothesis . . . to the general public without discussion or presentation. . . ."

Also in 1969, the Kanawha County (Charleston) Medical Society asked the state's medical association to oppose a drive for black lung to be listed as an occupational disease.

That same year Dr. Keith Morgan, director of the Appalachia Laboratory for Respiratory Diseases at Morgantown, West Virginia, said that a U.S. Public Health Service study showed that less than 3 percent of that state's miners had black lung "to a disabling degree."

Whether or not that was true, in March of 1969, the state legislature debated whether to enact a black lung compensation law. Just to give them a little guidance, Arnold Miller led 42,000 miners out the mines—most of them marching down Kanawha Boulevard in Charleston, to remind their state legislators that, at least once in their careers, their actions were being carefully watched.

Within twenty-four hours of Arnold Miller organizing a wildcat strike that would cost the operators $1 million per day, the legislators passed a black lung compensation bill.

Later in 1969, the U.S. Congress passed the Coal Mine

Health and Safety Act, which included national black lung benefits for the first time. Dr. Rasmussen and others complained that the federal bill contained X-ray requirements that could not detect most cases of black lung. After three years of pressure, the bill was broadened in 1972 to the extent that a great percentage of miners may one day receive compensation. The bill also raises future hopes of asbestos and cotton-mill workers for receiving similar compensation.

With the coal operators scheduled to pay the bulk of the black lung compensation by 1973, it was not surprising that they put pressure on the federal government to improve air standards in the mines. Thus an eager young deputy director of the U.S. Bureau of Mines, Donald Schlick, has claimed that his efforts would improve conditions so greatly that the next generation of miners might never have to worry about black lung.

"There's no way you can make the air safe," Isaac Reid says in his soft monotone. "You've still got to breathe. That 1969 act sets up a new ventilation system, but it don't make it better. Those air vents just recirculate all that bad air. They'll never make a safe coal mine."

Dan Sizemore stirs in his chair.

"Yeah, there could be," he says. "The Europeans use drilling and cutting machines that spray a watery substance on the coal while they cut it. We could use them, too, but it would cut down on the profit for the company. The big corporations would rather waste our lives than their money, boys. It's the same old shit. Profit rules everything."

Jimmy Reid peers out of his thin little glasses and talks in his mild, matter-of-fact way.

"We're a poor country in some ways. We've got a high standard of living but we're a poor country. How come the European countries take better care of their miners than we do? How come they had a law in 1934? How come we have to produce our sixteen tons a day while in England they only produce around four tons per man? They don't do anything in England unless it's safe."

The men run out of talk. They have said it all before. They look at the clock next to the portrait of John Fitzgerald Kennedy and they realize it's time for Dan and Jimmy to think about going to work. Isaac ambles out of the house and slowly works his way down the street. Dan gets in his car and heads for home.

"I don't tell many people," Dan says, after leaving Jimmy's house. "I don't want them to know over at the mine. But I went over to Doc Rasmussen's clinic last year to see if I could qualify for black lung. I took the tests with all those sad little miners. One guy vomited out his guts after walking on the treadmill machine. Another guy passed out after breathing into a sack, with some medic screaming in his ear to breathe harder. But that's the only way you can get qualified.

"Rass is a good man. I've worked with him on some things. He warned me he couldn't do any favors for me. Shit, I knew he wouldn't do anything different for me. He warned me not to have the tests sent to the company. They don't like their foremen to take the black lung test. He sent the tests to me personally. They showed I could retire on second-stage black lung. But what the hell. I still need the salary. Old Dan's got to keep going."

11

Dan can always smell the mood of the mine when he walks into the yard.

The mood today is not good.

The men are not tossing horseshoes and teasing each other. Instead, they are clustered in small groups, talking quietly. Dan sidles up to Frank Baker, one of his dearest friends, leaning against a pickup truck.

"They closed down the day shift," Frank says softly. "Inspector red-tagged Four South. Found some bad powder around. Sent the day shift home at noon. They're gonna have to get the powder out before we can go in."

Well, that should be no problem, Dan thinks to himself. At least, it wouldn't be a problem if we knew how to run a coal mine anymore. Estill Dean could pack the stuff carefully and bring it outside on the rock-dusting car. He could do it in an hour.

Dan drifts into the office and sees all the bosses from the day shift milling around, getting in each other's way. He assumes they are making plans for a rock-dust man to empty out the blasting powder, so production can resume. Dan stays in the corner, amused.

In the midst of the milling foremen is the man who stopped production—the inspector from the Mining Enforcement Safety Administration office over in Morgantown, a fellow named E. C. Ryan, wearing the blue and white federal overalls.

You didn't use to see as many federal inspectors when enforecment was in the hands of the Bureau of Mines, before the formation of MESA in May of 1973. Inspectors were a small bunch of former miners, hired by the government to keep production rolling without blowing up the mines.

Some were honest; others were the best friend an operator ever had. In some districts the federal inspectors used to call the operator up a day ahead of time, telling him what section they were going to inspect the next day. The operator would tidy up the section. He would warn his miners to bring chewing tobacco the next day, meaning they better not dare to smoke around the mines, which is the cardinal sin of mining. That was their clue that an inspector was coming.

But things got tighter a few years back. After the seventy-eight men were entombed in the Farmington mine over in West Virginia, even the U.S. Congress began to pay attention to the danger in the mines. A fearless former college professor named Ken Hechler, a critic of the Bureau of Mines and the UMW, used his position as Representative from Huntington, West Virginia, to push through the Coal Mine Health and Safety Act, codifying just about every detail of coal mining. President Nixon didn't want to sign it but the threat of wildcat strikes forced him to on December 30, 1969.

The act was controversial. Many coal operators said it made mines more dangerous than ever. Dan himself believes that it does not solve basic problems like ventilation and dust control.

But as the forceful Ken Hechler formed his bonds with other activists in the region—Doc Rasmussen and Arnold Miller, for example—they insisted that federal enforcement would prove the law was good.

As if to punctuate the dangers of mining, exactly one year to the day of President Nixon signing the bill, a doghole mine blew up in Hyden, Kentucky, killing thirty-eight men. Federal investigators said the crew had engaged in illegal blasting practices; the operator insisted the regulations of the 1969 act had contributed to the explosion.

As a result of Hyden, the number of inspectors was greatly increased. There were only 250 inspectors for 6,000 coal mines on the day the act was signed. By the end of 1973 there were 1,145 inspectors out in the field and dozens of young inspectors being trained at an academy in Beckley. The number of actual inspections had risen over 500 percent from 1969 to 1973.

Yet the inspectors were hampered because the top officials had never codified penalties for violations. And the penalties were supervised by a Republican fund raiser named Edwin Failor whose closest experience to mining had been as a traffic-court judge in Iowa. Millions of dollars of penalties were meted out—and later ruled illegal—during Failor's eighteen months with the bureau. Thus, everybody knew that the terms of the 1969 act were not enforceable by penalty but rather only by the presence of a Bureau of Mines inspector. Fortunately, more of them seem to be under pressure to make the mines safe.

Dan knows E. C. Ryan, and believes that E. C. Ryan is quite capable of closing this mine down forever—until the mine is safe.

Suddenly Dan realizes that something else is wrong. E. C. Ryan, who is normally the mildest of men, is turning red in the face and wagging his finger at the day-shift bosses. Ryan says, "This section is still closed and it won't open until we find the powder. And that's it."

And he stalks off to a telephone, to report to his district director.

Dan stumbles out of the office, into the silent work yard. He sees Frank Baker and Calvin Brooks and tells them the news.

"Boys, the powder's done disappeared," Dan says laconically. "They went in there and there wasn't a goddamn piece of it left. Now you know that our beloved gob rats may eat our lunch scraps, but they will not eat powder. That's a proven fact. So it must have been something human that moved that powder. And Big Ridge is going to be hurting until they find it."

Frank and Calvin and Dan look at each other with pained eyes. A deteriorating powder box could break apart and spill loose powder around a section. One loose spark could touch off an explosion and you'd have another Farmington or Hyden on your hands.

Or else Big Ridge will lose another shift of production. The way the company has been running the mine in the last few months, the miners dread every lost shift. They have heard the rumors that Big Ridge is planning to close down the mines around Milan. The men have seen the company investigators come snooping around.

Just a few months ago the company hired an efficiency expert from the steel industry, to discover why the company was retrieving only 40 percent of the coal instead of 60 percent. Some of the miners told the expert that the company was being run clumsily by John Drago, who was too busy romancing his secretary to concentrate on details. The efficiency expert must have made some kind of report. He was never seen again and John Drago is rumored to be headed for an early retirement. But the men at Big Ridge are not eager to follow him into retirement.

Dan wanders back inside the office. Things are quiet except for Larry Harding, the day-shift superintendent, who is obviously agitated about something. Dan watches Harding closely. Harding is one of the top bosses on the day shift, a former officer in World War II whose life reached a peak while he and a few million other soldiers chased the remnants of Hitler's army around Europe. He has much less enthusiasm for everything he has done since, including bossing the day shift.

"They ain't gonna find that goddamn stuff," Harding says darkly. "No goddamn inspector's gonna find that powder."

Dan stares at the super.

"Where's the powder, Larry?" Dan says softly.

"I hid it from the bastards," Harding says. "That's none of their business. Damn federal people gonna run our whole mine if we don't stop them."

Dan feels his stomach start to churn. He knows that if he blows his top, he'll have to fight the man, make an enemy forever. Dan knows all about mountain pride. He's played both sides of that game before. He knows if you cross the line with a mountain man, you've got to fight. There's no negotiating. So the trick is to talk sense to a man without crossing the line.

"Larry, I know how you feel," Dan starts slowly. "But this is gonna fuck us up. That inspector is gonna bring us more visitors. They'll have a team of those bastards in here tomorrow. They'll have a team of Big Ridge inspectors along with them. And you know as well as me that section has bad ribs and may not stand a close inspection. What do we want a whole bunch of them people coming around? If we don't find that powder, we'll have half the goddamn coal industry in here."

Harding doesn't answer. But he is listening.

"Larry, why don't we offer to make another search of the section," Dan says, extra casually. "Nobody has to know anything."

Harding still doesn't say anything. His lower lip is stuck

about six inches past his upper lip. But silently he follows Dan into the main office.

"Maybe we'll take another crack at finding the powder," Dan says to E. C. Ryan.

They round up Estill Dean and a couple of laborers and they pull on their helmets and set off in a man-trip, clackety-clack into the tunnel, to Four South, where the powder disappeared. There is a rectangular imprint on the dusty floor, where a box recently stood.

"Let's look this way," Harding says, playing the game.

They move down the shaft a few hundred feet to a cross-cut—a side tunnel, where a roof fall kept them from working. A huge pile of rocks and dust blocks the floor of the side tunnel. There are fresh footprints in the dust. The men climb gingerly over the pile, like boys exploring an excavation site after the workmen have gone home. On the other side of the pile they find the powder box, lopsided and old, a trickle of the low-grade explosive powder falling on the rocky floor.

"Let's get it out of here," E. C. Ryan says grimly.

Slim, immaculate Dean squats down and packs the old box in sawdust. Then he and the crew lift it cautiously into a larger wooden box. The three men handle the box easily. But whoever moved it originally must have struggled mightily over the pile of debris. The crew moves the box onto the man-trip. Dan and Harding and the inspector walk behind, nobody saying a word. E. C. Ryan does not ask why it was moved—or who moved it. But when he gets back to the office, he meticulously makes out his report, holding up the night shift until he has crossed every T.

The day-shift bosses leave, no questions asked. Harding stalks off silently, as Dan wonders whether the man feels that Dan has privately humiliated him.

At seven o'clock E. C. Ryan finally makes his telephoned report to his district office and gives permission for the mine to reopen. Dan knows he cannot work a full production schedule tonight, so he sends half his men home. Under union rules, they must be paid—since they were not notified before coming to work—but there is no sense in sending some of them inside tonight.

Some of the shift is kept around, including Calvin Brooks's production crew. Calvin shakes his head when Dan tells him the story of the lost and found powder.

"Dan, it sounds like children," he says in his full, deep voice.

"Absolutely," Dan says. "That man just lost his cool. I hate to see my side acting like that. Calvin, I swear, nobody has any control anymore. We resent the inspectors coming around. Why? Because the operators, the big bosses, keep telling us how lousy the law is, how lousy the inspections are. I swear, the company creates a man like Larry Harding. They tell their foremen not to discuss anything new at safety meetings. They just want us to accept everything. But look what they get themselves."

"Children," Calvin says, shaking his head.

Like prize athletes who have been kept on the bench by a rainstorm, Calvin and his crew rush into their section. With his continuous mining machine responding to his touch, Calvin moves nearly one hundred tons of coal in the three working hours available to him.

But the company has lost perhaps one thousand tons today, worth at least ten dollars per ton, maybe more. (The company does not ever say how much it receives per ton.)

But it is easy to figure that the night-shift salary cost Big Ridge at least $3,000 tonight. For nothing.

When Dan leaves the mine after midnight, he feels like crying.

Wednesday

12

Dan is working at the dining-room table, writing a speech he will deliver next month at Virginia Tech, to a group of engineering students who try to see past their professors on government contracts.

Just before noon, Dan hears a car motor outside. He looks out and sees a gray pickup truck climbing the dirt hill, stopping at the end of the path. The driver stares impassively as the passenger door opens and a forlorn figure trudges up the grass.

The driver waits until Edward has reached the steps, then he backs the truck around and drives away. The lettering on the side of the truck says "BRADSHAW COUNTY SCHOOLS."

Head hanging, long white hair almost obscuring his long white face, Edward shuffles into the house, casts his eyes on the floor, and waits.

"Well, your record is intact," Dan says flatly. "Every day this week."

Silence.

"Were you sick again?"

The pale specter nods.

"The secretary wanted to give me some milk of magnesia." The voice comes hesitantly. "But I told her that'd make me sicker. She gave me some aspirins but I didn't feel any better."

Dan nods momentously, weighing the knowns and the unknowns. Edward was probably up until eleven or twelve last night, until he and Gene dropped from exhaustion. This morning Edward probably didn't eat any breakfast before he sleepwalked down to the bus stop.

"Would you like some eggs and toast?" Dan asks. "David made more than I could finish."

The ghost looks upward through his white hair, nodding tentatively.

In the kitchen Edward starts slowly with some milk and toast. Then he shifts into gear with a dish of apple sauce and a plate of half-warm eggs. Then he takes a banana from the counter and some cold ham from the icebox, the remnant of last night's supper. The color runs back into his face.

"Reckon you'll survive?" Dan asks, drawing out the syllables.

Edward smiles, white teeth showing.

"Why don't you go up and take a nap," Dan suggests.

Edward bounces up the stairs. Safe.

Dan sits back at the dining-room table. Poor old Edward. Him and school haven't agreed since two years ago, when they kicked out the three boys for wearing their hair long. Dan and Margaret had to get a Civil Liberties lawyer down in Tri-Cities to take the case up to the district judge before they got a ruling that the school board could not dictate hair length.

When the judge ordered the board to let the three Sizemore boys back into school, the boys reacted in three different ways. Bobby just decided they were all a bunch of assholes and voted to stay away forever. Cocky little Gene thought it was great fun to swagger back into class and play king-of-the-school for the next few weeks. And Edward trudged back to class, dreading the attention he received. When people made a fuss over the hair incident, Edward just retreated inside himself.

Edward has always been the shy one of the family, even more than Mary-Ann. Dan thinks he has gotten more uptight since Margaret started school. On days when Margaret is not going to college, Edward can't be moved out to school. It's always the sensitive ones who suffer, Dan thinks to himself.

The boy is eleven now. He has been timid ever since he pulled that pot of boiling water on himself when he was three, scalding himself from shoulders to toe, turning red like some meat you cook. Dan got a friend to drive with him to the hospital in Tri-Cities, the friend wrapping towels around the boy, Dan barreling like a madman on the narrow roads, the smell of Edward's cooked flesh making him retch.

The doctor fixed up the burns as best he could, although some of the scars are still visible on Edward's chest and arms. But the doctor said he wasn't worried about the scars outside. He was more worried about the scars inside. Dan knows now what the doctor was talking about.

Still, Edward's a good boy, generous and bright. He'll pull through this, Dan thinks to himself. It would help if he had a good teacher some year, a teacher who didn't make school seem like a jail.

Upstairs, David's rock music is booming. But when Dan peers into Edward's room, the pale boy sleeps deeply, a big smile on his face.

13

The state of Virginia sits like a rusted wedge, misshapen and fuzzy, with its charming peninsulas and barrier beaches flaking eastward into the ocean; its tidewater cities clustering in the south; Richmond in the power base; Alexandria and Arlington in the populous northeast.

Then the state tapers down—Charlottesville with her historic university; the Blue Ridge running northeast to southwest; Roanoke ringed by mountains; the long ridges that channeled the pioneers to the Cumberland Gap, where Virginia comes to an end, farther west than Columbus, Ohio, or Detroit, Michigan.

It is doubtful if any other portions of American states are more cut off from their roots than Southwest Virginia. The mountains serve as a natural barrier. They also serve as a political barrier, since many mountaineers had no need for slaves in the early 1800s and therefore remained loyal to the Union and President Lincoln and the Republican party. In later years, when the South formed a bulwark of the Democratic party, the leaders in Virginia felt no compulsion to serve the western handful of counties, with their Republican roots.

Dan Sizemore once entertained several pre-law students from the University of Virginia, who were exploring the western portion of their state. One alert young man, a veteran of civil rights struggles in the state, confessed he had never realized that Virginia was a coal-producing state or had mountains other than the Blue Ridge. It is a well-kept secret in Richmond. Dan Sizemore insists that the late Senator Harry Byrd believed the state of Virginia terminated at Roanoke.

Yet some of the nation's mightiest corporations have discovered Southwest Virginia. They have built their mines and their freight lines and shipped the coal eastward to Norfolk or Baltimore. Today a solid portion of that coal is shipped to West Germany and Japan, a touch of irony that is not lost on men who fought in World War II and who hear the nation's leaders proclaim an "energy crisis."

The corporations that extract the coal from the mountains do not put much back into the region. A series of underassessments and tax loopholes make the Appalachian states a friendly harbor for the coal corporations. In neighboring West Virginia, for example, the state's main industry accounts for only 3.8 percent of the state's tax revenue.

The major loophole for the coal operator is the depletion allowance, which allows him to subtract the cost of mining the coal, preparing it, and transporting it up to fifty miles. Then he can subtract 10 percent depletion allowance before he figures his taxes. Sometimes he just *figures* his taxes; he doesn't necessarily *pay* taxes. In 1971 according to figures listed in the United Mine Workers Journal, Continental Oil—which owns Consolidation Coal Company—paid no federal income tax on

$109,000,000 income. The other major corporations do nearly as well.

The low assessments are tragic for the residents of the coal sections, because low state taxes mean poor roads and poor schools and poor health facilities. Since the Appalachian coal region is balkanized into Ohio, Pennsylvania, Kentucky, Tennessee, Alabama, and Virginia—only West Virginia can be considered a truly mountain state—the decisions for state tax allocations are made in lowland capitals, by legislatures dominated by farm and city interests and coal lobbyists. All coal regions are colonies; Southwest Virginia just happens to be more isolated than most.

This is where the Sizemores settled in 1958, after Dan Sizemore had gone without work for nine months. To locate the section of the state, it is best to start with the more familiar cities like Lexington, Kentucky; Charleston, West Virginia; Roanoke, Virginia; and Knoxville, Tennessee. Southwest Virginia is just about in the middle of them all, closer to other state capitals than to Richmond.

Nevertheless, new people have found Southwest Virginia, including a small wave of Italian immigrants early in the twentieth century, adding a few foreign names to the map or the phone book. The coal camp of Milan, whose apparent Italian origin has remained obscure, has long since been pronounced "MY-len." And a surprising number of men with Italian names have risen to prominence in several of the major coal companies, despite the general dominance by Scotch-Irish and English heritage. Few people think of John Drago, the boss of the Big Ridge operations, as being Italian. He has been around a long time. And everybody knows "Eye-talians" are outsiders.

The Sizemores first settled in Blake County, where the coal veins trickle out and the hilly farms predominate. It is a pretty country, but the Sizemores felt out of place among the farmers, who seemed secretive and greedy after the casual live-for-today mood of the coal camps.

After a year or so, the Sizemores moved up to Milan, where the Big Ridge railroad cars stack up for miles in the narrow valley, waiting to be filled with seventy or a hundred tons of coal. The steeper hollows, unfit for railroad tracks, were left to the humans.

The family felt more secure in Milan, living in a hilly camp. But when Edward and Gene came along, raising the family to eight children, the Sizemores found a bigger home a few miles away at the end of a dirt road, with four bedrooms and several huge rooms downstairs, the lovely front porch, and plenty of space outside for the children. Rent was only forty dollars a month; lately it has risen to fifty dollars. They have never owned a home.

When they moved to the house in the grove, the Sizemores unpacked—physically and emotionally. Big Ridge said the coal would last a lifetime. It was a good place to stop and discover how the fifties had changed their lives.

For one thing, the fifties had given them a new religion. Back in West Virginia, they had converted to Roman Catholicism, which was no small thing for a mountain family to do. Dan's mother had been a staunch Methodist with gentle prejudices against "coloreds" and Catholics. Margaret's folks were from the Community or Baptist churches, which regarded a loyalty to the Pope with slightly more tolerance than loyalty to Satan himself.

Both of them had known towns with ordinances against building a Catholic church within town limits.

Dan had known an old miner named Hoskins who told him that in 1928, when Al Smith was running for President, the people of Hoskins's village had gotten the word that the Pope was traveling through Kentucky, making whistle-stop speeches for Al Smith. Hoskins and his buddies went down to the station with a rope to hang the Pope, but when the daily train arrived from Cincinnati or wherever, there was no Pope.

"Well, Hoskins, did you really believe the Pope would come to your little town in the middle of nowhere?" Dan asked him.

"Dan, I carried the rope—I must have believed it," Hoskins said.

Although they were no strangers to anti-Catholicism, both Dan and Margaret converted to the alien religion. Dan said he had had enough of the Cold Creek Community Church, hearing the Reverend Billy Joe Hatfield comparing heaven to the way Joseph McGuire ran his company store. Dan said he felt the need for something better spiritually, with a sense of discipline and history rather than the emotional mountain religions, and the Catholic Church came up clean at that time. A trim well-spoken priest in Williamson reminded Dan of his father and helped convert both Dan and Margaret. He talked about having faith in the mystery of the Church. But later this priest left Williamson for a wealthier parish in Weirton.

When the Sizemores moved to Virginia, they joined the Catholic church near Milan, but they were immediately disappointed. Margaret began working with the young priest, doing research and editing his sermons. Dan began asking some questions about Pope Pius XII and his concern for the Jews during World War II. The priest didn't have many answers for them. After a while Dan felt that the priest, too, was pushing somebody else's company store.

They felt a stirring of pride in 1960 when handsome young Jack Kennedy campaigned in the West Virginia primary, insisting that a Catholic had as much right to be President as anybody. Kennedy seemed like a fresh breeze after the staleness of the 1950s. When he won the West Virginia primary that boosted him toward the nomination, the Sizemores celebrated for their country and their religion. But soon after that, they drifted away from the Church. The young priest persisted in paying visits to their home, sitting on the porch, arguing about salvation. Dan and Pete finally told him not to come around anymore. Chris, more withdrawn and gentle, continued to receive the sacraments for a time. But the family's affair with the Catholic Church was over. They had other things on their minds.

For Dan, the big old house gave him an opportunity to unpack some of the heavy books he had read when he was younger—the *Das Kapital* that used to belong to his brother Ford; the Dickens that told of the inhumanities in London; the Muckrakers of America. He began to read them again, putting them in the framework of the pink slip the Franklin Coal Company had handed him.

He sent away for more literature that described the relationship between capital and labor. He realized he had no faith in unions protecting the workers because he had seen John L. Lewis operate coal mines on the side, turn hundreds of thousands of miners loose from the union during hard times.

After Kennedy became President, Dan came in contact with the outsiders who began to stream into Appalachia with the VISTA and Appalachian Volunteer programs. Most of them were from northern cities, trained for a few weeks or months, then sent off to live in the smallest of rural villages. Many of them openly professed that they intended to help the people by changing their way of life.

The VISTA's and the Volunteers were gentle and kind toward the poor. They offered rides to doctors' offices; they provided baby-sitting care for harassed mothers; they brought down mimeograph machines and typewriters and showed some of the poor how to organize and publicize.

But the young outsiders were run out of more than one

town, because they offended the doctors and lawyers and coal operators and car dealers and school principals and undertakers who form the power structure. And they created no bonds with the working miner, who often resented their concern for the poor on welfare and the outsiders' long hair.

[Mountaineers all remember their grandfathers wearing longer hair with beards and mustaches. Yet, until recently, they adhered to the GI Joe crew cut, trying to assimilate.]

But while most miners and businessmen reacted angrily to the newcomers, Dan found himself listening to them. They acted as if they were living in a foreign country that needed changing. And Dan began to think to himself: Maybe they are right.

Before long, Dan began considering himself an Appalachian man, whereas he had always thought of himself as just being a typical hard-working American. He began to see himself as a minority citizen, one of the eight or ten million mountaineers scattered in the deepest hills from Western Pennsylvania down into North Georgia and Alabama. As he got more interested, he met some of the activists and teachers who were stirring in the early sixties—Philip Ware from North Carolina; Shirley Dittemore, a sociology teacher from Ashley Community College; and some of the Black Lung Association people. He also read the classic *Night Comes to the Cumberlands,* by a patrician Eastern Kentucky lawyer, Harry Caudill, a scholarly and human history of the colonized mountains.

As a result, Dan began to ask himself one of the toughest questions a man can ask—Who am I? He wanted to have the same pride as descendants of the Mayflower or fifth-generation Californians or the "Black Is Beautiful" negroes. But what were his roots? What had produced the tight little society dominated by a few coal companies and doctors and politicians?

Dan had always been taught the mountaineers were the direct descendants of the founding fathers, the hardiest sons who tramped off into the wilderness. But he began reading theories that the mountaineers were unskilled, uneducated ruffians who escaped bondage in the coastal states and opted for the anarchism of the mountains.

Harry Caudill, himself a descendant of an old mountain family, has suggested a great gap between the skills of the Seaboard colonists like Jefferson and Washington and the rough ways of the Spicers and Coxes who settled the hollows. Yet many people in Appalachia—particularly the townspeople—strongly resent the theory that schools and churches and even rudimentary house-building skills were late arrivals in the hills.

But Dan could see a certain sense to that. In his mind, if there ever was a truly free American, it must have been the early Appalachian settler. He was off in the mountains, hunting and fishing and growing his own crops, moving on when other families settled in his hollow. The forests were thick, the water was clean. In Dan's eyes, that early settler must have been king on earth.

But then came the land agents (Daniel Boone himself was an advance man for a land company). They bought vast tracts from anybody who would put an *X* on paper for fifty cents an acre. And then came the lumber companies, laying their rails, running the mountaineers off their property. After that, the coal companies, promising high wages, bringing in a cash society that trapped the mountaineer in a cycle of work, followed by a greater dependency on automobiles and power that strengthened the need for a salary. And then the whole region was hooked on working for the company. But the record clearly showed that the timber and the coal and the money flowed north and east. And somewhere along the line, too many people Dan knew had given up, had said, "Aw, to hell with it," and went on welfare and took the sedatives from the local medicaid doctors or accepted a stipend from the courthouse gang for doing some dirty little deed. The exploitation of the mountaineer was just about complete.

Although he had never joined a union, except for a brief time on a construction crew, Dan had once thought that unions might be an answer to the exploitation. But he felt that John L. Lewis had let the miners down. This was brought home again in the sixties, when the impoverished UMW closed a chain of hospitals it had established, leaving the miners and their families with no source of health care. In the long run, who would stand up for the mountaineers? What was the answer for America's poor?

Dan Sizemore arrived at the answer on his own. Nobody brainwashed him; nobody forced him into it. Certainly peer pressure had nothing to do with it. He decided that the profit system was the enemy. The fact that men in New York made a profit had forced Dan Sizemore to be a bully foreman in unsafe mines.

Strictly on his own, Dan decided that no large industry should be privately owned. Any shop or business that employed more than a couple of workers should be owned by the state. He started reading about depletion allowances for the coal company, getting madder than hell that he could not take money off his income tax for his blackened lungs and his weary body. He decided that competition between companies had ruined his lungs. He decided it would be better to spread out the work, three or four days a week guaranteed for everybody, than to have booms and busts, layoffs that crippled people forever. But he knew that capitalism would never reform itself.

So in his own words, "I guess you'd say I became a Marxist. But I was no Commie. I was an American who didn't believe the profit system could work. That was all."

Of course, that was when his anger was mostly theoretical. That was before Vietnam.

14

The argument is raging in the lamp house when Dan arrives. The voices bounce off the low ceiling; the wooden benches scrape against the cement floor. Dan stands in the doorway and listens.

"I ain't gonna work with no damn woman," shouts Darrell Kinzer, a slender, middle-aged roof-bolter with a red-lined drinker's face. "This job is tough enough for a man without no damn women around."

Paul Grose, a chubby powder man with a graying beard, shouts back at Kinzer.

"You don't know these women," Grose says. "They could take it. They've got just as much nerve as any man, including me."

Well, the secret's out, Dan thinks to himself. One of the day supervisors had told him last week that four women have applied at the personnel office over in Milan. It didn't take long for word to pass around. But then again, just about everybody has a relative or friend in Milan. Grose—who is part of a small Hungarian family that settled here a long time ago—is cousin to one of the women.

"How is a woman gonna like working in our mine?" asks Deacon Lambert in his gentle, near-lisping voice. "We don't have any sanitary stations for women. Women are used to things being nice."

Not only that, Dan thinks, but women are supposed to be one of the worst jinxes inside a mine. Every time a mine blows up, people wonder if a woman was inside. Over in Hyden, when the Finley mine blew up in 1970, all the widows kept saying it was because the owner used to bring in nurses from the Frontier Nursing Service, just down Hurricane Creek. The federal people later blamed the explosion on the miners using illegal explosives. But that didn't change the superstition.

Even disasters like a stillborn child are sometimes attributed to women visiting the inside of a mine.

All of this is strictly a recent American tradition, of course, since women used to work alongside men in British and Eastern European mines. Women have also worked in small "family" mines in the United States. But no woman has ever belonged to the United Mine Workers or worked in any of the doghole companies that anybody knows about.

Yet women visit mines all the time, superstition or not. Margaret has visited mines with her father and also with Dan. Dan remembers bringing Mary-Ann over to the mine when she was around seven or eight, but she got so frightened when she saw Dan in his dirty clothes that she ran crying to the car. She has never gone back to a mine since. Vicky, on the other hand, rides over whenever Dan's friends come to visit him at work. Vicky even talks about working in the mines when she gets older.

The men are accustomed to Dan's long-hair and no-bra

"outsider" friends driving up to the mine. One time a tough little Welfare Rights Organization gal from over in Morgantown came to visit the mine. Big, slow Elmer Spicer offered her a bite of his lunch, just to show how hospitable he was. She wolfed down an entire sandwich and would have gone right through the rest of the lunch, too, just for laughs. The other men didn't stop laughing for a week.

Then there was the time Shirley Dittemore came over. Shirley is no stranger to the mines. This soft-eyed sociology teacher from North Georgia has been researching mining for over a decade without causing any mines to blow up. But the men didn't know that at first. All they saw was a gentle lady with the graceful tones of the South.

Lawson Blair, one of the shuttle-car operators, asked her if she felt she could handle a tough job like working in the mines. Shirley took a look at his relatively clean uniform, deduced that he was no mechanic or roof-bolter or any other dirty specialty, and replied that some jobs might be too tough for her—but she guessed as how she could probably do some of the easy jobs, like running a shuttle car. Blair nearly had a stroke, the men laughed so hard. He might have gone away with hard feelings for all womankind if Shirley had not favored him with one of her soft smiles.

But Vicky and the WRO gal and Shirley Dittemore are only visitors. A woman working inside the mine would be a different thing, Dan figures.

The men are so used to their own company that any intrusion, any change could affect their daily routine. Perhaps they would be more attentive, more productive. Or perhaps they would be so distracted that they would fail to pay attention to danger signals inside a mine. And any kind of jealousy or shift in friendships could imperil the feeling of trust that enables a man to work in the mine. Although Dan knows it doesn't always work out that way, most miners feel secure that their "buddies" will make any sacrifice to help them in an emergency.

Over the years, the men build up friendships with each other that are not equaled in many marriages or families—although most of them would be embarrassed and would deny it if the subject were verbalized. Men call each other "buddy" and "honey" with an open affection that seems more European than American—and quite foreign to a visitor from the cold, impersonal urban world. Their horseplay, on the other hand, is rougher and cruder than anything seen in most factories or offices.

The men call young recruits "boy" or "son." And it is always a shock for a middle-aged miner the first time somebody calls him "Uncle"—the half-mocking nickname for anybody older. Estill Dean never infuriates Dan more than when he calls him "Uncle."

Dan thinks: If women came to work in our mine, what would happen to the grand old tradition of "hairing"? This treatment is for a miner who has annoyed his fellow workers, usually for some personal habit like farting or strewing garbage at the dinner hole rather than for bad working habits. The offender is seized by several men who yank off his pants and pull out his crotch hairs, by the roots. In some mines this is a standard form of initiation for a new miner. Sometimes it leads to vicious fights. But usually it is accepted as the way things are done.

Having women in the mines would also cut down on the men's sex lives, Dan thinks. Most miners are capable of topping their women eight or nine times a night, according to their own testimony. However, these estimates might drop considerably with a woman working alongside a man.

"I don't think they can take it," Darrell Kinzer continues. "They'd get in some tight spot with the foreman down on their ass and they'd say, 'I don't have to do this. I can always wash dishes in a restaurant or go on welfare.' Hell, a man's got to work till he dies. A woman don't have to work. It makes a difference, buddy."

Dan stares at Kinzer. Did women drive him to drink or did drinking give him his low opinion of women? Dan has no idea.

"Listen, I know two of these women," Paul Grose comes back. "They're kin to me. They've mucked out stalls all night after working in a sewing factory all day. Hey, this would be like a vacation for them."

Raymond Wilcox, the man with the 1950s crew cut and the body that has been broken so often, nods his head in agreement.

"I don't know how I'd feel about it," he said. "But geez, women used to work as riveters and welders during World War II. They did all right, didn't they? Some people told me the women worked better than the men because they were neater."

"A woman couldn't take it," Kinzer says, his voice growing harsh.

The way Dan hears it, Grose's two kinfolk are mother and daughter. The older woman is Liz Dickerson, a wiry grandmother of around fifty years of age. Her husband used to be a miner until he lost his leg in an auto accident. Mrs. Dickerson has run the house and the farm since her husband was disabled—while she was working full-time in one of the nonunion factories over in Morgantown.

Because the area has generally been ignored by the International Ladies' Garment Workers Union, the companies have been able to get away with all kinds of dirty tricks on women. When Mrs. Dickerson went out sick with the flu, after not missing a day of work in five years, she came back to find her wages had been dropped from $2.50 to $1.60 an hour. She argued with the supervisor and then just walked out.

That night, according to the story, Mrs. Dickerson was home watching one of the talk shows and she heard one of those women's libbers from New York, Gloria Steinem or one of them, talking about how women could do anything a man could do. The idea just clicked in Mrs. Dickerson's mind. Why should she struggle in a sewing factory for $1.60 an hour when men were making $5 an hour in the coal mines?

A few days later, Mrs. Dickerson and her daughter and two single women appeared at the Big Ridge office, asking for job application papers.

Apparently one of the secretaries smarted off at Mrs. Dickerson. Dan can imagine the scene: a snippy townie from Milan, her hair all spun up in the 1950s beauty-parlor look of middle-class mountain women, talking down to these four ridge women in their overalls or print dresses. At any rate, the four women started yelling back, until the personnel manager, William Foster, escorted them into his office. It took Foster fifteen minutes to realize the women were serious about working in the mine. He told the women that legally he could not deny them a job on the basis of sex. But he also told them Big Ridge was currently laying men off, rather than hiring. He said they could fill out their applications in case Big Ridge ever started hiring again.

Now that's a cheering thought, Dan thinks. But what if the price of coal goes up again—and the company decides to expand its crews—and the women force the company to hire them? Dan only hopes he is still bossing when the time comes.

"Suppose you got some nice-looking woman down in the mine," one of the men says. "Everybody would be trying to big her instead of doing their job."

"You couldn't even tell the difference between a man and a woman in our uniforms," another man says.

Estill Dean slips in at the fringes and says, "The Bible says in Genesis that woman is made for childbirth and a man earns his living by the sweat of his body."

"Yeah, but this is a different time," Paul Grose says. "Women have proven they can do it. Look, Dean, you and me work on the same powder car. What's the toughest thing we do? Lift a fifty-pound case of powder? You see women mowing the lawn or moving furniture. They can move a sack of rock dust, too."

"Well, I don't know," says Deacon Lambert. "I guess if they came around, I'd try to warn them about the dangers. I'd try to help them."

"I'd quit first," says Darrell Kinzer.

Dan feels the argument is starting to die. He stirs himself to get dressed.

"You fellers are forgetting one thing," he announces, pausing in the doorway. "Anything nice comes on our shift, I'm gonna break her in first. Wouldn't want some poor woman to learn mining from a bunch of raw laborers like you."

"That's just like you, Dan Sizemore," one man retorts. "Hog it all for yourself. Don't share nothing."

Dan waves and does a fast jitter-step out of the lamp house.

Thursday

15

Margaret has no school today. The family is holding a minor reunion in the kitchen. At ten o'clock Margaret is cooking a big bunch of eggs and bacon for Dan and David and also for the two youngest boys, Edward and Gene, who are now both suffering from the same invisible malady that has plagued Edward all week. The pale boy rarely makes an attempt to go to school when his mother is home these days. He stays with her, from room to room, secure and happy. Gene, who hates to be left out, generally stays home on Thursdays, also.

Gene is the last of the Sizemores, a pretty boy nearing his second decade, with a round baby face, ginger-colored hair, and a family reputation for exaggeration. He is an extrovert, the other side of the moon to Edward.

David is the most talkative this morning, reviewing his plans for getting a minibike next Christmas. When he gets the bike, he insists, he is going to run over Mrs. Bailey on the lower end of the grove, because he thinks Mrs. Bailey shot his dog, Blackie, two years ago. Somebody shot Blackie.

Also, David plans to visit the mine, as Dan promised.

"Daddy's taking me to the mine," David says. "When you gonna take me to the mine, Daddy? You gonna take me tomorrow?"

Dan is enjoying this midweek gathering. He has already shaved and dressed in clean clothes.

"David, now I can't take you on a weekday," Dan says. "However, I promised you a ride on the man-trip and Saturday's the day. Two days from now, David, unless they run a full shift, which I doubt."

Margaret bustles over the hot stove, aware of performing a familiar task that has slackened somewhat since she started college. In all her years of making breakfasts, Margaret never questioned being home to serve her husband. It still seems the right thing to do; it is going to college that gives her twinges of guilt.

But neither was she the traditional coal miner's wife, who knows only what her husband chooses to tell her about his work. Dan and Margaret talked openly about his job from the first. When something was bothering him, she felt she had a right to know. She would ask him why certain things were done at the mines. Sometimes they would argue—good strong arguments, with him losing his temper and launching into lamp-house language, but the talking and the arguing led to a deeper bond.

Some coal marriages do not seem so open. After the Hyden disaster on December 30, 1970, many of the thirty-eight widows came forth to testify that their husbands had alluded to strange practices at work.

"My Alfred used to lie in bed at night and say, 'They're gonna kill us all if they don't stop what they're doing,' " one widow told a panel of investigators.

But the widow had apparently never felt inclined to question her husband about what "they" were doing—which, for the last desperate weeks before the explosion, seemed to be using outdoor, road-construction "prima-cord," a high-intensity explosive guaranteed to spark off the volatile coal dust in a doghole mine.

It was the ultimate Catch-22 of coal mining:

—The miners hinted at dangerous practices.

—The widows didn't ask questions.

—Asking questions might have made the husbands quit a risky mine.

—And then the family would have been broke.

—Therefore, if the wife used her sense, she'd take food off the table.

All this happened, of course, without their realizing the vicious cycle of their unasked questions. But that has never been a problem for Margaret Sizemore. The woman who once sat by the window with a rifle and vowed to risk her life against a gang of thugs has no fears of understanding her husband's dangers. And when Dan had come home with his face bloodied from the deputy pistol-whipping him, it was Margaret who nursed him and helped give them courage to stick it out in Bradshaw County. So the details of the ramshackle Big Ridge mines hold no new terrors for Margaret Sizemore.

Breakfast is over now. They sit in the living room and lazily smoke and drink coffee. Gene has discovered a box full of color snapshots. He lays them on the coffee table and tells the story of each photo.

"That's when I caught the sunfish and Pete and Chris didn't catch anything," Gene says, delighted with himself.

"They were too busy baiting your hook," Dan says.

Gene shuffles through the pictures. He uncovers a picture

of a lovely young girl wearing a T-shirt, the rain making the T-shirt cling to her breasts.

"How come Pete split up with Jenny?" Gene asks.

"Oh, they had an argument, I guess," Margaret says, absentmindedly. Jenny had lived with Pete for almost two years after his marriage broke up.

The pictures are from the Sizemores' vacation in Toronto last year. Every summer the coal industry shuts down for the first two weeks in July and miners roam far from the Appalachians, seeking clean streams and unmarked landscapes. They drive to Florida to the beaches or Michigan to the lakes, anywhere to escape their destroyed homeland. The Sizemores cross the border into Canada, to see their two oldest boys.

They leave on Saturday morning, all eight of them piling into the family sedan, which Margaret packed the night before. They barrel north, stopping only for gas, with Dan refusing to permit anybody to touch the wheel. His temper grows short when he approaches the complex loops around the big cities; he curses the Interstate Highway system as it cunningly plots to confuse him.

Late that night they arrive in Toronto, at the high-rise where the two boys share an apartment. They stay up late that first night, trading stories of bungling coal superintendents against bungling civil service chiefs (both Pete and Chris have work permits and are employed by the government).

As the dawn approaches, Dan and his two sons grow more hysterical, laughing ecstatically over each wisecrack. One time Dan read a short story he had written, about a pet dog that got crushed by a pet horse. The three of them laughed so hard that Margaret stalked out of the living room, complaining that all three of them were heartless and cruel. But the laughter finally subsides; they sleep late on Sunday; then they start a whirl of sightseeing and fishing and gorging themselves in Toronto's inexpensive Chinese and Italian restaurants. And then the two weeks are over. On the last Sunday morning, the family piles back into the sedan and says good-bye to the boys until Christmas.

Last summer they rode across the flatlands of Michigan and Ohio, caught between their boys and their mountains. But magically in the dark of evening, just as they crossed the Ohio River into West Virginia, the car radio played the new song that has become almost a state anthem: "Country Roads." Dan insisted everybody wake up and sing along; the magical coincidence was too much to ignore.

Then they slept again while Dan burrowed deeper in to the valleys of his native state, heading for the distant corner of Virginia they had adopted, thinking of the time when there were eight children at home, and wondering whether he would have to quit mining before Nixon would give amnesty, to put the family back together again.

16

Pete was the first to question Vietnam, in the year he turned eighteen and Lyndon Johnson was elected as the peace candidate.

Pete had always been an independent boy, full of ideas, full of fight. He made up his mind early that the war was worthless and he announced that he would not serve. Period.

Dan and Margaret argued with their oldest son for a while. They had been raised in another era, in a region that produced the highest proportion of Medal of Honor winners, amidst a feeling that a man always "did his duty" for his country.

Even though Dan never joined the service, even though he was beginning to admire certain aspects of socialist economy, he still supported Johnson and the war effort, partially out of fear of Goldwater.

But in 1965 Dan was working the hoot-owl shift and he came home from work just as the Fulbright hearings went on television. He would postpone going to bed just to hear Dean Rusk and some of the others try to explain the war. The more Dean Rusk talked, the more Dan turned against the war. It just didn't seem to make sense.

By the end of 1965 it was Chris against his brother and his parents. The sensitive second child argued that the United States was fighting against atheistic Communism.

The Sizemores, who were still toying with Catholicism at the time, went to the young priest down in Milan and asked his opinion on Vietnam. When he waxed enthusiastic about stopping Communism, they recalled his vague explanation of Pope Pius XII and the German-Jewish question. So the three of them stopped going to church completely.

One day Chris blurted out that "the three of you are destroying all hopes of being with God in eternity." He still received the sacraments—"the last of the Catholic hard-hats," in Dan's words.

Once he turned eighteen, Pete had to register for the draft. But he could avoid the army for the time being if he attended college. Pete, who had never been an eager student, started over at Ashley Community College, studying engineering. But his mind wasn't on college. His mind was on Grace Ratliff, who lived over in Morgantown and looked older than her fifteen years.

Dan and Margaret tried to talk their son out of getting serious with Grace. They told him that if he didn't concentrate on college, his only future would be the army and the mines, the traditional route for most mountain boys. But Pete snapped back that college was just a status symbol for his parents.

In later years, Margaret would agree that Pete might have been right. But she defended pushing college by citing the tough time Dan had encountered without any college education, particularly in mining subjects.

At the start of the second semester, Dan and Margaret gave Pete $215 for tuition and books. Instead, he and Grace took the money to North Carolina and got married. When their money ran out, they came to live with the Sizemores.

With no job and no school, Pete was just marking time until the inevitable letter from the draft board. There was no such thing as draft counseling in Southwest Virginia. Margaret got the idea that the navy was better than other branches since it was not as heavily involved in Vietnam. She called down to Tri-Cities and a navy recruiter came up to visit.

But when the recruiter heard that Pete had been convicted of a misdemeanor for "vulgarity" in high school (he said he had merely propositioned a girl who was eminently propositionable), the recruiter said that Pete would not be accepted as a naval volunteer. Pete said the man had wasted his trip anyway; he wasn't about to join any branch of Lyndon Johnson's armed services.

The letter came early in June of 1966, ordering Pete to report for induction on June 13 in Roanoke. On the day before he was due, Pete and Grace threw their clothes in a suitcase, the boy not quite twenty, the girl not quite sixteen.

Dan stayed outside while they packed, raking leaves furiously. He watched them hail down the afternoon bus, going north.

When Pete and Grace got to Toronto, they wandered around the terminal until they saw a poster from the Students Union for Political Action, offering counseling. SUPA got them a room and advised them how to emigrate legally. Pete sent a telegram home, asking for $200 so he could cross the border to Syracuse and return as an emigrant. His parents also had to send a telegram to Canadian officials promising to support the couple until they found work.

Margaret worried all night, positive the FBI would arrest

the couple at the border. But on Friday Pete called to say he had been accepted into Canada.

For four months the Sizemores supported the young couple, even allowing the company to repossess their new Lincoln, saving them $100 a month in payments. The family had never been without a car before and they felt isolated at the top of the grove. But the savings meant that Pete and Grace could exist in Toronto until SUPA found Pete a job.

Then it was Chris's turn. He was heading toward eighteen in 1966, as Lyndon Johnson kept promising about lights at the end of the tunnel. Chris was beginning to waver about the Church; he was also beginning to doubt the war. But he said it was wrong to go over the border, the way Pete had done. Chris vowed he would go to prison first.

This put Margaret—as the miners say—between the rock and the hard place. If Chris served his time in prison, he'd be legally free to stay in the country. But she had read about how draft-dodgers were treated in prison, often left at the mercy of the rugged older prisoners who often were homosexuals. She was afraid that Chris would not be able to survive the brutal prison system. After all the vicious arguments she had had with Pete, Margaret had vowed never to interfere in her children's lives again. But she told Chris she would ask only one more thing out of him the rest of his life—that he not go to prison, when his time came. Chris finally agreed to cross the border instead of going to prison. But Margaret still thinks he felt guilty about going to Canada.

After his brother left, Chris seemed to change. He had always been the neatest of the children, collars always even, shirts and socks always fresh. But now he began wearing sneakers without socks, an old leather jacket he had found in a rummage sale. People said Chris was looking like a hippie. Margaret wondered if he was trying to hurt her.

Chris went up to visit Pete around August of 1967. On his way back in the fall, he stopped at a demonstration at the

Pentagon and was arrested. He stayed in jail several days rather than post bond. Later he told his parents he had chosen jail to see if he could take it.

In the spring of 1968 the army sent its notice for Chris. On the chilly day when Martin Luther King was buried, the boy tossed some extra clothing into a tote bag, said good-bye to his parents, and walked down the hill. He said he didn't want them to know where he was going–so they wouldn't have to lie to anybody. A few days later, he called from Toronto to say he was living with Pete and Grace.

Now the Sizemores began to catch hell from two sides–their family and their government.

Almost all their relatives disapproved of what Pete and Chris had done and they stopped communicating with the Sizemores. It wasn't until 1971 that Dan would discover that one of his nephews had been killed in Vietnam. His brother had not wanted Dan coming up to West Virginia for the funeral.

One of Dan's sisters called him on the phone and asked him, "Why are your sons draft-dodgers?"

Dan replied, "You've been teaching Sunday school for thirty-seven years and you have to ask me that?"

He would not hear from his sister for four years.

The chief support in the family came from Margaret's mother, who now lived by herself down in North Carolina. While the rest of the families cut off contact, she continued to call and write. For that gesture of family loyalty, she was rewarded by harassment from the FBI.

One FBI agent called up the old lady and asked why she had not visited Margaret lately. The old lady said that's just the way it happened, no particular reason. The FBI agent said, "You know what could happen to you if you don't tell the truth?" Mrs. Hamilton was so afraid that the FBI would badger her for the boys' address that she refused to write it down. She was afraid the FBI would catch her in some kind of lie.

The Sizemores wondered how the FBI had located Mrs. Hamilton in the first place. Their only conclusion was that the agents were tapping their phones or stealing their mail—or both. Margaret used to laugh at herself for being so paranoid about the chief investigative body in the United States. Dan was not so trusting. He said he was sure Johnson and later Nixon would permit all sorts of spying to harass antiwar people.

Sometimes the Sizemores would hear a click on their phone, nothing like the party-line sounds. Other times they'd hear machines humming, sounds of an office they'd never gotten from the telephone company.

Other times the FBI would openly call or even visit the house. Some somber-voiced man would say he was just checking to see if the boys had sneaked home during the night.

"They know they can't find Pete and Chris," Dan warned Margaret. "But they're just hassling us to scare other kids, like Bobby and some of his buddies. They want to keep the next wave of kids frightened."

Grace's family also received visits from the FBI agents. It began to unsettle the Ratliff family, who had not taken kindly to Pete evading the draft. It didn't take long before Mrs. Ratliff began putting pressure on Grace to leave Pete and come home.

After one year in Toronto, Grace came home to visit her family. She was only supposed to stay two weeks. But Mrs. Ratliff destroyed Pete's letters and refused to let Grace talk to him on the phone. Grace found out what her mother was doing and called Margaret. Margaret told Grace she was a married woman now, who could do anything she wanted. If Grace wanted to return to Toronto, Margaret said she would send a taxi over with enough money for a bus ticket to Toronto.

Grace went back. Soon she was pregnant and delivered a

boy named Pete Sizemore, Jr. Mrs. Ratliff visited Toronto and got in a shouting match with Pete, who kicked her out of the apartment.

Shortly after having the baby, Grace got pregnant again. Margaret often wondered if Pete had deliberately given Grace two babies quickly, to induce her to stay.

When Grace was three months pregnant, Pete came home from work one day and discovered Grace and the baby were gone. The landlady said she had seen Grace's sister hanging around for the past few days but she hadn't thought anything of it.

A few days later, Grace and her sister and little Pete returned to Morgantown.

In 1970 Grace gave birth to her second boy, Andrew. She also got a divorce and married a boy from Ashley County.

Grace's new husband is fond of Andrew but refuses to raise little Pete, who lives with Mrs. Ratliff.

Once in a while, the Sizemores see Grace while shopping in Morgantown. They visit their grandchildren on rare occasions, when Grace's husband isn't around.

People in Morgantown say the husband is jealous of Pete and refers to him as a "chickenshit draft-dodger." People also say that Grace always speaks nicely about Pete.

17

"Hey, Dan, come over here and look at this."

Dan never knows what to expect when he enters the lamp house at 3 P.M. One day it might be Tierney, the young powder-car man, showing the latest Polaroid pictures of all the college girls he gets to pose in the nude. Another day it might be somebody with a fishing story or a hunting story.

But today it looks like Grant Harris, holding a handwritten letter, flanked by Jimmy Reid and Jack Reese. Looks like a chapter meeting of the Miners for Democracy, Dan thinks as he strolls over. Three of the leaders when we got rid of Tony Boyle as president of the United Mine Workers.

Dan hits the mark exactly. Grant has a letter from a cousin over in West Virginia, telling them about a group of disabled miners and widows. Angered at being denied medical cards because they didn't have a full twenty years of union experience, they had helped turn their southern West Virginia county against Boyle in the election of 1972.

"My cousin says these folks can't get their medical cards under Arnold Miller, either," Grant says, his thick face turning red. "Union says they can't afford to change the rules just now."

Anger shows on Jack Reese's bladelike face.

"That was one of his big promises in the campaign," Reese snaps.

"That's what I mean," Grant says. "Them politicians are all the same. Once you elect them, they forget about you. Looks like Miller's done the same to us."

Dan feels the other miners in the lamp house staring at him, waiting for his comment. It is no secret that Dan was an ardent supporter of Arnold Miller, just like this trio of miners. He made speeches for Miller, he housed several of the youthful college students who organized the district. But as a foreman who has never belonged to the United Mine Workers, Dan has to be careful at the mine not to seem too partisan.

"Well, he's only had a few months," Dan says. "You can't imagine Miller to clean up ten years of Tough Tony in only a few months. It's the same thing as mining coal. You've got to have a plan."

The trio of men look at Dan with blank faces. They have heard his bitter tirades against Tony Boyle, away from the mine. They have heard Dan praise Arnold Miller as a man of the people. They know his position as a foreman; but somehow they expect more of Dan Sizemore.

The men are quick to feel betrayed, Dan thinks. That is the other side of the coin to their loyalty. The poor devils would cut coal with their foreheads if you asked them to. Most of them probably risked their lives in one of our wars, just because some dumb officer waved the flag. But then when it's over, the poor miner feels betrayed.

It's a problem in the mountains—just as with the blacks. There just isn't that much money or power around and when somebody gets ahead, the resentments build up. Either he misuses his new power—or the men suspect him of misusing it. Either way, a new demon emerges. And the men search for a new hero.

The loyalty-betrayal syndrome is only an exaggeration of the national character that leads Americans to seek easy answers. If only a few hundred prisoners of war are returned, everything will have a meaning. So they put bumper stickers on their cars and they despise anybody who does not give three cheers for POW's and MIA's. We're a nation of easy answers, Dan thinks.

But then he realizes that miners have good reasons for feeling betrayed. Almost from the beginning, they've had valid reasons for thinking somebody was out to get them. In the early days it was the Baldwin Felts detectives with their machine guns and the National Guard at Ludlow, Colorado, and General Billy Mitchell flying his planes into West Virginia to break up a strike. These heroes of capitalism made the miners loyal to the United Mine Workers. They held secret meetings, they braved the company thugs. And when they were recognized in the coal boom of the twenties, they felt a great surge of loyalty to John L. Lewis. When the union built hospitals and passed out medical cards generously, the men felt their prayers had been answered.

But later the men learned how John L. Lewis was running a nonunion coal company on the side. And when the profits sagged and the companies failed to keep up their payments to the union treasury, the network of medical care collapsed. Then the union cut off membership in Eastern Kentucky and elsewhere, dooming the men to doghole standards.

The isolation of the coal region kept most men in ignorance, however. If angry miners felt the union was not backing them up at their mine, they might stage a wildcat strike. But miners fifty miles away would never hear about it—and often the striking miners would not see the overall pattern.

For example, when miners began questioning their lung troubles, they got little help from their local doctors or union officials. The majority of newspapers gave prominent space to the spokesmen for the major medical societies, who denied that black lung existed. And the most popular source of news for the coal miners, the union's *Mine Workers Journal*, was content to run as many photographs of the miners as it could squeeze in, on the theory that a man would rather see his own face than read the bad news about safety conditions or pulmonary ailments.

The leadership of the union passed in 1959 from John L. Lewis to his assistant Tom Kennedy, and then to Lewis's assistant W. A. (Tony) Boyle, in 1962. Boyle knew how to keep the union in line. His loyalists were rewarded with local positions worth fifty dollars and higher per month. They were given trips to "safety committee" meetings in the bigger cities. Union officials frequently disappeared during the summer months on business trips and came home laden with treasures. Estill Dean returned from one "business trip" with two luscious salmon from the icy Michigan lakes. He gave one to Calvin Brooks and another to Dan Sizemore, knowing their taste for tangy foods. Dan thinks: How can you despise a man who gives you a whole, fresh salmon?

Still, a reform movement emerged in 1964 and Tony Boyle dealt with it in the direct coal method at the 1964 convention. A squad of Boyle loyalists, wearing white helmets, was organized from the District 19 delegation from Southeast Kentucky and Eastern Tennessee. Whenever an anti-Boyle speaker would make a comment from the convention floor, somebody in a white helmet would hit him with a club. Many of the delegates seemed to accept that anybody who tried to question Tony Boyle deserved anything he got.

The miners who didn't attend the convention did not learn about the bloody white-helmet squad from their UMW *Journal.* Subsequent issues of the union journal contained pictures of happy old pensioners clutching their $150 monthly checks or photos of Tony Boyle greeting a safety committee—or recipes for corn fritters. Yet the UMW *Journal* was often the only reading material that a miner ever saw, while only those who subscribed to the coal region's few responsible newspapers like the *Louisville Courier-Journal,* had a chance of finding out what really happened.

The isolation in the United Mine Workers continued through most of the 1960s, even with the emergence of the Black Lung Association. And the isolation might have continued forever—except for a mysterious change of chemistry in a trusted UMW official, Joseph (Jock) Yablonski.

Jock Yablonski was a rough customer. He was a former miner from Clarksville, in the southwest corner of Pennsylvania. He had risen in the hierarchy for being tough and shrewd—and loyal. He had a comfortable income from the union and a secure future. But something happened in the 1960s that made Jock Yablonski change. Some people have suggested that putting two sons and a daughter through college had opened Jock's eyes to his role in the union. Others say that his wife, Margaret, a bright active woman who once had a play produced near their home, had raised his social conscience. It was much the same kind of middle-age change that Dan Sizemore went through.

Whatever the reason, after the dissent over Vietnam and the police riots in Chicago and the student rebellions, Jock Yablonski called a press conference early in 1969 and announced he was running for president of the union.

Yablonski accused Boyle of ignoring health and safety conditions, leading to disasters like the one in November of 1968, where 78 miners died at Farmington, West Virginia. He accused Boyle of mis-management of contracts and pension funds. He accused Boyle of secret deals with coal companies. And he suggested that he knew many more secrets about the leadership of the union. And with only a few trusted friends and limited finances, Jock Yablonski began touring the coal states, slipping into the bathhouses to make harsh speeches about change.

The miners at Big Ridge remember Jock Yablonski. At first Dan and a few others had informed the miners that Yablonski was running against Tony Boyle—something that was not always apparent in the UMW *Journal.* But the union loyalists had painted such a dreadful vision of this turncoat that the Big Ridge miners would not have been surprised to see Jock Yablonski arrive with horns and cloven feet—or on a leash of a coal operator.

He drove up the mountainside in the fall of 1969 and instinctively found the bathhouse. As the amazed, silent miners watched, Yablonski warned them of dirty business in the union hierarchy. He was thick and rugged; his eyes flashed and his meaty fists pounded the air; he seemed like a mountain preacher but his message was revolution rather than pie-in-the-sky. Some of the miners hissed "Commie" at him and hardened their faces or walked away. But other men's eyes filled with the knowledge that an insider was saying the same things they had always suspected about their union.

When Yablonski left that night, some men cursed him and others promised to contribute to his campaign.

Tony Boyle visited Big Ridge a few weeks later. Always poorly advised, like many leaders with entrenched power, Boyle had underestimated Yablonski's potential at first. But as the campaign continued, Boyle apparently realized that Yablonski was attracting attention wherever he went. So Tony

Boyle scheduled some trips to the bathhouse for himself.

A small, dapper man with thinning, red-gray hair, Boyle materialized at Big Ridge one afternoon, wearing brand-new overalls that looked as if they had just been purchased at the company store. He was also chewing tobacco, with a little trail of juice trickling down his pasty-white chin. Boyle was obviously trying to remind the men that he had once been a miner himself. (Or so he said. Although Boyle claims to have worked in his native Montana, records are spotty about his working career. Most of his adult life was spent as a union official while other members of his family were coal operators.)

The miners stared at the little old man and contrasted him with the booming Jock Yablonski. And while Boyle was reminding the miners about their duty to be loyal, Dan Sizemore gestured to Estill Dean to perform his favorite trick and goose Boyle from behind with his middle finger. But Dean, who held a minor role in the local union, shook his head and refused. (Dan told him later, "Estill Dean, if you ever goose me again I'll kill you, because you didn't have the guts to goose Tough Tony Boyle.")

However, another man did dare to take on Tony Boyle. Jack Reese, who had fond childhood memories of John L. Lewis visiting his father's farm on union business, felt that Tony Boyle had let the union down. As the president was shaking hands with the miners, Reese inched his way up to Boyle and said in poor-mouth fashion, "Mr. Boyle, it sure is nice to see you here. We kind of wish you could visit us more often."

Taken by surprise, Boyle launched into a solemn speech about "Now, son, you know that I have many pressing affairs back in Washington to take care of. If I spent all of my time with you boys down here . . ."

Reese suddenly lost his tone of servility. With knifelike bitterness, he interrupted Boyle.

"Well, then what in the goddamn hell are you doing down here today?" he asked.

Apparently shocked, Boyle stepped backward and looked around for Carson Hibbetts, the district president, who had driven him to the mine. But Hibbetts was unpopular at Big Ridge because he had supported the management when they suppressed a wildcat strike. Hibbetts was, in fact, hiding in his car behind the powder shop.

Finding himself standing alone, Boyle lost his composure. His pasty face reddened, he shook a few more hands, and he bolted downhill until he found Hibbetts's car. A few moments later, the two union officials went barreling down the dirt road, eager to get back to friendly territory again. Boyle never came back to Big Ridge.

The campaign was much more hazardous for Jock Yablonski. One night in Springfield, Illinois, he was struck from behind by a karate blow while attending a meeting organized by Boyle loyalists. He lost consciousness and barely managed to drag himself back to Clarksville.

The vote was held in early December. While the Yablonski supporters felt they had reached many miners during the campaign, they discovered that the union had been busily preparing its election strategy. On the day of the voting, Yablonski voters often could not discover where the balloting was to be held. Other times they would arrive and discover that the voting had been held earlier in the day. Although Yablonski's supporters had persistently requested government supervision of the election, the Department of Labor had steadfastly refused to look into union affairs. Yablonski's people therefore realized they had no way of knowing how many illegal votes were being tabulated. When the vote was announced by Boyle's headquarters, Boyle was the winner, 80,577 to 46,073. It was the closest vote for president in the history of the United Mine Workers.

According to the union's own tally, the Big Ridge miners had voted for Yablonski by a 3–2 margin.

Yablonski refused to give in, announcing that he was suing Boyle for illegal campaign practices. As he returned to Clarksville for the Christmas holidays, friends warned him not to relax his vigilance. A strange car with Ohio license plates had been seen cruising around Clarksville. One of Jock's friends did some investigating and found the plates were registered in Cleveland. He called the owner and was told by the wife that her husband was in Clarksville, looking for a job.

On the night of December 30, three thugs recruited from the hillbilly ghetto of Cleveland crept into the darkened Yablonski farmhouse and shot Jock Yablonski, his wife, and his daughter while they slept. Their bodies were found by their son nearly a week later. On the windowsill, a slip of paper contained the mysterious Ohio license number that Jock's friends had investigated.

Weary miners everywhere nodded to each other and said, well, that's what happens when you buck the union or the operators or the government. The rest of the country was just discovering violence in the 1960s. Hell, the miners had lived with it for fifty years.

Back in Pennsylvania, Jock Yablonski and his wife and daughter were buried on a January day so bitter that tears froze as they trickled down the cheeks of his friends. Two of Jock's closest supporters—Joseph Rauh, the founder of Americans for Democratic Action, and Mike Trbovich, the shuttle-car operator who had been Yablonski's campaign manager—met at the funeral and vowed to keep up Jock's struggle. They would call their group "Miners for Democracy."

Within two years, Miners for Democracy enrolled thousands of men, including Jimmy Reid, Grant Harris, Calvin Brooks, and Jack Reese over at Big Ridge. With Joe Rauh providing legal assistance in Washington, and with a handful of mountain radicals and young non-miners joining the struggle, they continued Jock Yablonski's fight.

The Miners for Democracy formed close bonds with the Black Lung Association and lobbied for the enforcement of the 1969 Coal Mine Health and Safety Act. They helped win a suit that eventually ordered four union officials to repay $11.5 million to the Welfare and Retirement Fund as a penalty for keeping pension money in noninterest accounts in the National Bank of Washington, which was owned by the union. Although not named in the civil trial, Tony Boyle was ordered to resign as a trustee of the fund.

The union was also convicted of conspiring with the massive Consolidation Coal Company to force the small, nonunion South-East Coal Company of Kentucky out of business. The UMW had to pay $4 million to SECO.

The presence of the MFD emboldened miners everywhere. At Big Ridge the men had grown tired of having their shifts changed or their jobs changed, whenever the company felt the need. They had found the district offices unwilling to take up their grievances. So they began responding with the traditional mining call for a wildcat strike—by emptying their water buckets and walking out of the mine. Usually Big Ridge would lose several production shifts before persuading the men to return.

The company finally obtained an injunction against the miners from walking out. Then the miners received a letter from the UMW's young lawyer, informing them that their personal belongings could be taken if the penalty injunction were enforced. This letter made many miners more bitter than ever, because it seemed that the young union lawyer was taking the side of the company and the government.

Dan teasingly reassured the men that he would gladly purchase their favorite motorcycles and fishing equipment at a reasonable fee if they were penalized for the wildcat. The

Miners for Democracy had gained a few more recruits.

The militance grew in the coalfields in 1971, as the three-year contract with the coal operators came to its end. Feeling pressure from the MFD, Tony Boyle staged a six-week strike for higher wages, even paying a total of $100 in strike benefits for the first time in union history.

When Boyle approved a contract in November calling for a $50 maximum daily wage, the miners approved the contract—but thousands of MFD loyalists stayed out of work another week, to voice unhappiness over the lack of fringe benefits. If Boyle had been paying attention—or if he had had good advisors—he would have seen that the MFD was making deep inroads into the rank and file.

But Boyle was too busy in court. The Department of Labor had been forced to get involved in union affairs after the deaths of the Yablonskis and was now joining the MFD in asking for a new election because of irregularities in the 1969 vote.

Also, Boyle was convicted in 1972 of illegally using union funds to contribute to political campaigns. He appealed a five-year jail sentence.

By 1972, the three killers of the Yablonskis had been caught and their trail had led to Silous Huddleston, a former local official from the dreaded District 19, who pleaded guilty in planning the action. Huddleston then implicated Albert Pass and William Prater, two of the top officials in District 19. A mysterious fund of $20,000 had apparently been used to finance the murders.

That money seemed to have come from two $10,000 checks authorized by Tony Boyle to District 19 in the fall of 1969. But Boyle insisted he had merely given Pass and Prater money for organizing in District 19, with no questions asked. After both Pass and Prater were indicted in the murder in 1972, the prosecuting attorney, Richard Sprague of Philadelphia, suggested bluntly that the orders had come from the very top of the union. Some people doubted if the murders would cost Tony Boyle his job—since coal miners had long become accustomed to violence.

In the spring of 1972 a federal judge in Washington, William Benson Bryant, overturned the election of 1969, ordering a new election by December of 1972, under strict supervision by the Department of Labor.

On the Decoration Day weekend, hundreds of MFD members gave up their traditional family homecomings and graveside services. Instead, they streamed into Wheeling, West Virginia, to nominate a successor to Jock Yablonski. At first it seemed that the intense Mike Trbovich would automatically assume the leadership of the insurgent group, after running the MFD for over two years. But perhaps Trbovich's moody ways or his unpronounceable Croatian name hurt him. After some backstage maneuvering, the nomination went to the hero of the Black Lung Association, Arnold Ray Miller from Ohley, West Virginia. Trbovich accepted the bitter blow and agreed to run for vice-president.

Just as Jock Yablonski had done in 1969, Arnold Miller went out to the bathhouses in 1972. But it was different this time. With the Department of Labor entering the case, the MFD was suddenly legitimate. Company officials usually made a pretext of politeness or, when they barred union speeches from their grounds, claimed to enforce the rule equally.

The biggest difference was with the miners. Where Yablonski had been unknown and suspected when he began campaigning, Arnold Miller's approach was familiar to the miners. The younger ones in particular—back from Vietnam, with longer hair and a sense of things wrong with their country and their union—occasionally gave Miller a fisted "power salute" when he entered their bathhouse.

Miller appeared at Big Ridge one afternoon wearing a gray suit and a plain tie. Not even the company executives who fly in by helicopter wear suits to the Big Ridge mine. But Miller was trying to implant an image of himself as a man who could run a union despite his limited education and lack of office experience. The suit and tie were part of the image.

The men looked him over, most of them for the first time. They saw a graying, square-faced man, one ear mangled from the machine-gun bullets at Normandy, his face impassive as he searched for words. The suit could not disguise Arnold Miller's origins. As he shifted his weight from one foot to the other, the men could sense that he would be much more comfortable in overalls and mining boots.

"I'm Arnold Miller and I need your help," he told the men, biting off his words nervously with huge gulps of air, so that each word was half-swallowed. "Biggest issue I can see is giving the union back to the men. Got to have representation. That's why I want to listen to you."

A dirty miner from the day shift barged into the bathhouse, eager to go home. When he spotted the crowd around Miller, he surged over to see, but he held back his grease-covered hand.

"That's all right," Arnold Miller said, extending his hand. "I'm used to dirt. When I was a repairman, my wife said I used to crawl in the stuff."

The miner shook hands with Miller. All around the room, the men seemed to smile. He was one of them.

Tony Boyle stayed away from most bathhouses in 1972, but he did hold a series of rallies every weekend. His kickoff rally was on Labor Day in Grundy, Virginia. Despite a hard rain, Margaret and the two girls drove several hours to hear Boyle speak. Dan figured he had better stay away rather than be seen at Boyle's rally.

Surrounded by hard-faced assistants, Boyle looked tiny and pasty-faced, his few strands of hair slicked on his head. But when he spoke, Boyle came alive with crackling anger. Staring down at the several thousand miners and their families standing in the rain, Boyle charged that the U.S. government was trying to break the union by promoting the three MFD candidates. He accused Joe Rauh of controlling the three candidates—Miller, Trbovich, and Harry Patrick, the candidate for secretary-treasurer. He called them "the Three Stooges." A low chuckle spread through the audience.

Boyle shouted how he had raised salaries and pensions during his ten years in office. He agreed that he had contributed to political campaigns—for the good of the union. And he said he would be proud to go to jail for fighting for his union. And he insisted that he had fought for health and safety and black lung benefits.

Many of the listeners cheered and applauded. Some of them had driven down from West Virginia to hear Boyle speak. And there was Estill Dean, wearing a spotless set of green overalls, cool and unruffled as always, talking quietly out of the side of his mouth.

"That's a fine man," Dean said. "He's done so much for us miners. It would be a damn shame if they didn't support him now. I can't see any reason for not voting for him. He's doubled my salary. That's all I know. I'll be making $50 a day in another year."

An old pensioner leaned against the fence of the high school football field and said, "Tony Boyle doubled my pension. It was $75 a month when he started and now it's $150. He'd do more if he could, too. How do I know these new fellows won't cut my pension. I'm sticking with what I know."

When his speech was over, Boyle smiled and shook hands with a swarm of admirers. Then he excused himself. He had to catch a plane back to Washington by Monday night. Under the court ruling, he had to be in his office on Tuesday morning, performing union duties, or he would not get paid.

The campaign went into its final months as the Appalachian fall hung mistily over the coalfields. Boyle based his entire campaign against the "outsiders," the people behind Arnold Miller. There was some basis for his charges. Joe Rauh was not from the coalfields and he had raised thousands of campaign dollars from wealthy liberal friends in the northeast. Boyle said that Rauh was hungry for a power base in Washington and that he was using "the Three Stooges" to further his goals. Many miners, with their history of betrayal, saw no reason not to believe Boyle.

While Boyle was tied down to Washington during the week, the three MFD candidates seemed to be everywhere, each of them visiting a few bathhouses every day. Remembering what had happened to Jock Yablonski in 1969, they stayed away from strangers as much as possible, traveling by car, staying in the homes of trusted friends whenever they could. One night in a hotel in Whitesburg, Kentucky, two men with Harlan County license plates insisted they had to see Arnold Miller at three in the morning. But the hotel clerk had been tipped for just such an emergency and he insisted that no Arnold Miller was registered.

When the candidates visited Pennsylvania, they stayed at the rural cabin of Ed Monborne, their national campaign chairman, whose wife plied them with steaks and pies until they could eat no more. Or back in Clarksville, Mike Trbovich's wife would dole out sausage and cottage-fried potatoes until the candidates stumbled out to rock on the porch, too full to move. Then, in the long evenings, miners would drop by the front porch and talk over the campaign.

Yet the Miners for Democracy were not all miners. College

students and former VISTA volunteers, the radical youth of the mountains, began flocking to the cause, setting up arrangements for the eight days of voting in December.

The Sizemores housed a young law student named Harold Sexton for over two weeks around election time. He slept until noon and made telephone calls for much of the day, cajoling people into organizing car pools and poll-watching committees. Other times, Harold would talk Vicky's and Mary-Ann's boyfriends into giving him rides to distant mines or meeting halls.

Harold had never worked in a mine, yet he was very quick to give orders to the miners. When Dan heard Harold on the phone, laying down the law to some fifty-year-old black lunger, Dan winced with pain for the man's mountain pride. Yet Harold seemed unaware that his methods might annoy the mountaineers.

"I'm driving them ruthlessly," he announced to Dan one afternoon after a three-hour session on the telephone.

The Sizemores were almost ready to throw Harold out of the house by the time the election started. Later they said that Harold, in his own way, was working for the same goals they were.

The election began on December 1, in bitter cold throughout most of the coalfields. With hundreds of Labor Department officials watching every single polling place, the voting went smoothly for eight days. Then the ballots were taken under heavy guard by chartered bus, truck, and airplane to the government offices in Silver Springs, Maryland.

On December 16 the Department of Labor announced the results.

Arnold Miller had 70,373 votes, Tony Boyle had 56,334.

On December 22, as around three hundred refugees from the coalfields jammed into tiny McPherson Square in Washington, D.C., Arnold Miller was inaugurated as president of the UMW.

On September 6, 1973, Tony Boyle was arrested and charged with murder in connection with the Yablonski deaths. On September 24, the night before he was due in court for a hearing, Boyle took a heavy dose of sleeping pills and nearly died. However, he survived to face future trial for murder and the impending sentence for illegal political contributions.

In Miller's first few months, he fired as many Boyle loyalists as he thought necessary, he auctioned off the union's fleet of Cadillacs, he ended the union's association with the National Bank, he made plans to move union offices to West Virginia. He also ushered in a new period of interunion activism by supporting hospital strikes in Kentucky and Virginia and a garbage workers' strike in Charleston. He also supported the first major test of his administration—a strike against a nonunion company in "Bloody Harlan" County.

And when a fatal explosion took place in a Consolidation Coal Company mine just before New Year's Day, Miller chartered a plane in bad flying weather, to investigate the accident and to deliver a stinging warning to John Corcoran, the president of the massive company, that conditions had better improve—quickly.

Yet Miller's loyalists also saw some discouraging signs, when a few Boyle loyalists were allowed to remain in union posts. Also, when miners began running wildcat strikes again, Miller urged them to follow the proper grievance procedures before they walked out. Used to long delays and lack of support from the union in the past, some miners grumbled that Miller seemed no more militant than Tony Boyle ever was.

And when people reminded Miller of his promises to raise

pensions and reissue hospital cards to those who had been cut off, the word from union headquarters was that the treasury was worse than anybody had expected and that no new benefits could be issued immediately.

Dan Sizemore stands at the edge of the miners and listens to their grumblings. He knows that the men have felt let down so often that it might take a superman to satisfy them anymore. He hopes Miller is close to a superman.

18

Vicky rushes upstairs to change out of her dress, impatient to put school behind her. She wears cut-off dungarees and a sweater and nothing on her feet.

"Do you want to shoot the rifle?" she asks in her deep, slow voice. "We bought some shells last Saturday."

Mary-Ann says she doesn't want to shoot the rifle, it's too noisy, but she will take a walk up Milan Mountain, where the late-afternoon sun is still shining from the opposite ridge.

We come out the back door, climb the steep grassy lawn, scramble up a dry creek bed by holding on to the banches of trees. Vicky is carrying the rifle. Then we reach a dirt road, about ten feet wide, that follows the contour of the mountain.

This is the strip road, where the trucks carried the coal off the mountain a few years ago. The tire tracks are about six inches from the edge. Every few yards, the outer lip of the road has been tamped down with coal slag, to prevent wash-outs.

We climb and climb. Then we are halfway up, looking straight down at the Sizemore house, looking tiny now. Just below us is a huge mound of coal slag, piled against the mountain. This is the "red dog" the Sizemores discovered after they moved in. The first time it rained, a flood of ashes poured into their backyard and they raked it out for days afterward.

A few scraggly bushes and trees are trying to grow on the pile below us. Up above, the hillside is littered with boulders and broken tree trunks.

"This mountain used to be covered with trees," Vicky says. "Right over there, we hung a tire on a rope from a thick branch. We could swing right out over the hollow, like Tarzan. Me and Bobby used to feel like we could fly. They knocked that tree down when they stripped the mountain.

"In the wintertime, we used to sleigh-ride if we had any snow. You can't sleigh-ride anymore because of all the boulders they knocked down. You'd kill yourself."

We climb to the top of the strip road, where it levels out, a flat bench, fifty feet wide, dotted with pools of stagnant water, scrubby grass, discarded machine parts, beer cans, and bulldozer ruts. It looks like the moon, with garbage.

On the inside of the "bench" there is a vertical "highwall" of mountain, about fifty feet high, all dirt and rocks, scraped bare when they took out the seam of coal.

Above this obscene right-angle cut the mountain foolishly resumes its angular ascent for another hundred yards—aged pine trees crowding each other, unaware of the vicious surgery that was performed below. But at the bottom fringe of the slope, the dirt has washed away, uprooting trees.

Vicky picks up a section of cardboard box. She runs ahead and hangs it from the branch of a withered tree. She comes back and fires the rifle at the makeshift target. The cardboard jolts slightly after the rifle goes off. When I shoot, the cardboard droops undisturbed. We tease Mary-Ann into squeezing off one round. She closes her eyes and grimaces and pulls the trigger, missing everything. Then she gingerly hands the rifle back to Vicky.

"It hurts my ears," Mary-Ann says, almost apologetically.

The Sizemores remember worse concussions than the rifle noise. They remember the way the mountain looked when they first moved here, all lush and green. But then the red dog pile sifted into their yard, reminding them of the coal that always seemed to follow them.

Shortly after moving in, Dan tramped through the woods and discovered concrete stanchions in the undergrowth, part of an old coal tipple from thirty or forty years before. He realized there had once been a small underground mine somewhere in the valley. And they had dumped the waste higher on the mountain, a typical practice in the coal region.

Then in 1969 surveyors started cutting their way up the hill. The man who rented them the home, Mr. Cooper, said he was leasing the upper portion of the mountain to a local strip miner but, he continued, the Sizemores were welcome to continue renting down below.

Up until that point, strip mining had always been something that happened to other people in faraway lands—like leprosy. The papers had carried stories for the previous five years about occasional gun battles in Kentucky, where mountaineers chased strippers off some hill. Once in a while the Sizemores would see a mountain being stripped in some other county. But they always considered it somebody else's problem, a local aberration. Surely nobody would want stripping in Bradshaw County.

The strippers came one morning and bulldozed a road along the most shallow contour of the hill. They crunched huge trees into the Sizemores' backyard, tumbled boulders within a dozen yards of the house. When they reached the seam of coal that ran the length of Milan Mountain, they started trucking the dynamite up the hill.

The first blast came early one morning while Margaret was fixing breakfast. People always assume that noise is the worst aspect of dynamite—but it isn't. The worst part is the concussion, the wave of power that hits you deep in your belly, that makes dishes clatter and windows shake, the whole house creaking in protest.

And then another blast a few minutes later. And another. The blasting went on for days and weeks, until the Sizemores considered leaving their solitary outpost. But finally the strippers had exposed the seam of coal and the explosive charges grew smaller, more localized, as the derricks loaded the loosened coal into huge trucks that grunted their way down the steep strip road, their air brakes gasping as the drivers maneuvered, a few inches away from tumbling, crashing death. Each gasp of the air brakes sent tremors into the Sizemore house below. But the coal got out. The coal always gets out. And the only thing hurt was the mountain. When the strippers moved down the ridge, over another hollow, they left behind a trail of rubble. Mr. Cooper, the owner of the land, said it didn't bother him any. It wasn't disturbing the Sizemores' house, was it? Well, then, what were a few uprooted trees, anyway?

Dan carried the inquiry one step further. He telephoned to Richmond, to the state agency in charge of strip mining. The man said the job had been successfully completed in the eyes of the state. The company had sowed grass seed and carted away the coal. The bond had been returned to the company. Maybe the law wasn't perfect, the agent allowed, but the company hadn't done any different than all the other strippers.

But with the trees gone, and the new grass insufficient, every time it rained now, the mountain couldn't hold the water. It poured off the hill, washing dirt and boulders with it, disturbing the pile of red dog worse than ever. After a while, the Sizemores gave up complaining about it.

What happened above the Sizemores' heads was nothing unusual. Strip mining has gashed its mark onto thousands of miles of mountaintops wherever there is coal. In recent years, Piedmont Airlines, which serves the deepest portions of Appalachia, has received occasional complaints from passengers who are appalled at the endless vista of mountaintop scars. Older passengers take it all in stride. They can even knowledgeably point out the narrow old strip mines in contrast to the wide new cuts made by modern machines.

The coal region was never exactly noted for its conservation practices, not in a nation notorious for being unable to cope with its garbage. Since they first started mining, companies have been dumping their residue in the nearest hollow, just like the company that left the red dog above the Sizemore house. And the coal companies have traditionally used the broader streams to dispose of their acid runoffs. When they dumped their burning coke alongside the road, where it smoldered for months, pouring acrid smoke into the noses of motorists, that was considered normal, too. And when the tipples and cleaning plants released dust and smoke into the air, dirtying laundry and reddening eyes for miles, that was considered normal as well. But the mess always seemed localized, concentrated, near the mine shafts. Back up in the hills, where the true mountaineers lived, it was assumed that life would go on indefinitely, with people planting vegetables, fishing from the higher streams, taking drinking water from the creeks, burying their dead on sunny plots facing the east.

Then World War II ended and American technology turned to "peaceful" ventures. One of the targets was the earth itself. Huge earth-moving machines were created that made it profitable to scoop the surface dirt away, to dig dozens of feet down to the seams of coal, rather than burrow elaborately in the traditional deep-mining fashion.

It started as a simple exercise in isolated corners. But bigger and better bulldozers started coming around steeper and taller mountains. And then the boulders started tumbling into the cornfields and creekbeds of Appalachia, and the water started to taste bad and children started to get stomach cramps from well water. . . .

Because mountaineers were traditionally isolated, it took them a long time to realize what was happening to them. If a man in Jones Creek saw the strip miners working above his home, he approached it as his particular problem; he did not usually bother to call fellow mountaineers in neighboring Smith Creek. Sometimes he tried to fight it; often he moved; but usually he assumed, in fatalistic mountain fashion, that what was happening to him was unique. And if it was happening elsewhere, it didn't concern him.

So, without any opposition, coal operators pursued their technological breakthrough into the early 1960s as if they were exploring a new frontier—which, in a sense, they were. The days were long gone when a rugged individualist could ride through the West, laying claim to vast tracts of land. But in the mountains of Appalachia any man who had the guts to put a down payment on a D-9 bulldozer had a very good chance of making a fortune.

It was like Sutter's Mill, all over again. Miners who had feared they were destined to spend their lives underground, working for some giant corporation, suddenly became opportunistic employers, working in God's own sunlight, turning over more coal and more money in a few hours than they used to see in weeks.

Profits for strip miners have been estimated at 126 percent of investment. Stripped coal can be produced at a cost of fifty cents per ton, while deep-mined coal is estimated at $2.75 per ton.

The strippers began driving new Cadillacs or Oldsmobiles instead of Fords. They began moving to prosperous mountain towns like Pikeville, Kentucky, which boasts of having over fifty millionaires—more than one millionaire per hundred residents. The wives of strip miners began visiting the hairdresser two or three times a week instead of the mandatory once, their hair done up in the elaborate twisting fashions that are a status symbol in middle-sized mountain towns. Strip miners began running for state legislatures, got invited to be trustees for colleges, built country clubs, and dug swimming pools.

With the coal region balkanized among eight Appalachian states, most state legislatures treated strip mining as a distant industry that produced revenue for the state but affected few of the state's residents. The state governments passed out permits for strip jobs as if they were fishing licenses. Their rule seemed to be: if a man dared to work on a steep hill, it was legal. Reclamation was usually defined as removing all workable equipment and hauling away all salable coal. The strippers often elected one of their own to represent them in Frankfort, Richmond, Nashville, or Charleston. He would tell the urban state legislators that everything was just fine back home. Usually he would get appointed to the legislative mining committee.

After this diligent pioneering by the early strip miners, the big energy corporations moved into strip mining in the middle 1960s. They started to buy out the small operators, making them richer, so they could move away to Florida and tell stories about the miserable hillbillies who didn't have enough pride to clean up their creeks.

The 1960s were almost half gone before the people began to realize what was happening.

The first screams of outrage came from the hills of Eastern Kentucky, not because the mountaineers there were any more enlightened but because the laws were particularly favorable to stripping in that commonwealth, whose laws are patterned more on French law than on English.

Kentucky is a strange state, caught in a twilight zone between north and south, between east and midwest, unsure of its character and drifting in the fog like a modern-day Brigadoon.

By the middle 1960s, only in Kentucky were citizens in danger of being stripped on their own land. This was because of a legal device called a "broad-form deed," a contract signed generations ago by unsuspecting mountaineers, selling their "underground" mineral rights to land companies for fifty cents or a dollar per acre.

While the original intent had been to deep-mine the land, when strip mining grew into a common practice, the land companies often invoked the "broad-form deed" and stripped somebody else's farmland. In most states the public outrage was so great that the law was gradually refined to make strippers negotiate a new deal with the landowner. But not in Kentucky. To this day, the holder of a broad-form deed signed three generations ago can legally strip somebody else's land.

Not even the isolation of Eastern Kentucky could prevent the horror stories from spreading in the 1960s. People told stories of cornfields torn up, creekbeds flooded with silt, barns knocked over, outhouses destroyed. One woman named Ritchie watched with horror as the stripper followed the seam of coal straight into the family graveyard, tumbling the coffins of several of her dead infants into the lower cornfield.

Accompanying these stories of uninvited stripping were tales of Kentucky justice. In one courtroom after another, Kentucky judges ruled that the broad-form deed was legal. More than one judge sat behind his bench with his boots still muddied from a morning of surveying his own strip-mining operations.

Although the companies made few efforts to contain the damage or to reclaim the land, they occasionally cleared away a landslide they had created. In the predawn darkness of Knott County, the school bus would often run into a landslide caused by rains of the previous night. Then the children would wait on the narrow road until the strip miner gallantly dispatched his bulldozer and cleared a path.

Some mountaineers accepted the stripping stoically, as an indication of the hard life the Lord had intended for them on earth.

Other mountaineers felt the Lord had meant for them to fight back.

In Knott County, a couple of proud old-timers, with the courage from a disappearing age, decided they were not going to be pushed around.

In the fall of 1965, Mrs. Ollie Combs, a sixty-one-year-old widow, sat down in the path of the bulldozers that were stripping her family land. The sheriff came and put her in jail overnight for creating a disturbance, but the national press and television picked up the story and made a heroine out of "the Widder Combs."

Not too far from the Combs land, Uncle Dan Gibson was managing several farms for his nephews who were off doing their duty in Vietnam. Uncle Dan was well into his seventies; he was a preacher who kept his hunting rifle well oiled. The way Uncle Dan figured it, if his nephews could fight for freedom in Vietnam, he could certainly fight for freedom in Knott County.

When the strippers started to move onto Gibson property, Uncle Dan put his rifle to his shoulder and began drawing his sight on the lead bulldozer. It took a squad of state troopers to talk the preacher into putting his rifle down before somebody got hurt. And the only way they did it was by getting the stripper to promise to move his operation elsewhere.

That was in 1965. Uncle Dan delivered the land intact when his nephews came home.

Word of Uncle Dan's victory spread around Knott County

until other people decided to emulate his stand. One foggy night a few patriotic law-abiding citizens disabled a $20,000 Euclid earth-mover by firing a bullet through its radiator. From the cover of thick trees, they told the bulldozer operator to go back to wherever he came from. He went.

This particular guerrilla war was aimed at Bill Sturgill, a beefy ex-athlete from the University of Kentucky who had a long-term contract with the Tennessee Valley Authority that was making him rich. Sturgill wisely moved his operation from this hot area and people began priming the Widder Combs and Uncle Dan for sainthood—the Betsy Ross and George Washington of Knott County. But it didn't work out that way.

Bill Sturgill found another hollow ten miles away, where residents were not prepared to fight, and he stripped so much of Knott County, leaving the hollows so torn and desolate, that not even a hermit would live in some places. Jean Ritchie, the noted folk singer, does a song about her native Knott County. The song is titled "Black Water."

Bill Sturgill later sold his company for a $1 million profit to a huge power syndicate from Houston. Other new power companies, with no roots in the mountains, had much less compassion about stripping than the old pioneers. The power companies hired mean-eyed foremen and distributed pistols and whiskey whenever there was a demonstration against stripping. The guards did not hesitate to bash heads; the anti-stripping tactics became more legal than physical, for good reason. Naturally the police looked the other way when strippers were chasing demonstrators away.

It was impossible to hide the destruction of the strippers, particularly with papers like the *Louisville Courier-Journal,* the *Nashville Tennessean* and the *Charleston Gazette* beginning to educate their urban readers. Under some slight moral, middle-class pressure, the state legislatures gradually passed strip-mining laws. But the committees were usually dominated by strip-mining agents and the resulting laws—except in

Pennsylvania—generally did more to legitimize stripping than to regulate it. In Kentucky some of the strongest critics of the state law were often the few conscientious agents who tried to enforce it.

Meanwhile, technology outraced the imaginations of the legislators or the worst dreams of the mountain residents. The Peabody Coal Company built a machine called "Big Muskie" —ten stories high, as wide as an eight-lane highway—that tore up an entire county in Western Kentucky. And the Hanna Coal Company unleashed a monster called "the Gem of Egypt" (*G*iant *E*arth *M*over in the Egypt Valley of Southeast Ohio), that weighed fourteen million tons, stood 170 feet high, and could gobble 200 tons of earth per bite.

Operating twenty-four hours a day in plain view of Interstate 70, where gaping tourists watched this glittering dinosaur rampaging in the darkness, the Gem created ghost towns, turned the countryside to rubble, and induced more than one nervous breakdown by its destruction.

With giants like the Gem and Big Muskie, stripping became an equal partner in the coal industry. By 1972, 52 percent of all coal was mined from the surface. One government estimate said that 20,000 miles of scars were left in the eight Appalachian states.

But the most fertile coalfields were still barely touched. According to one official survey, 77 percent of all coal that could be mined economically lies west of the Mississippi River. By the 1970s, the broad-form deed had popped up in such non-Appalachian states as Montana and Wyoming, where ranchers had signed away mineral rights in earlier generations, never dreaming that huge veins of coal would one day be discovered under their fragile grazing topsoil.

Railroad and power companies began purchasing huge tracts of land from any ranchers who felt the need to sell out. And they prepared to invoke the broad-form deed against the proud sheep or cattle ranchers who had always imagined themselves as secure as any person in America. But with their mineral rights sold out in broad-form deeds, these hardy ranchers began to see themselves as helpless as any poor mountaineer.

As the stripping moved westward, even the Congress of the United States began to get involved. Coal mining had traditionally been the responsibility of the Interior committees, whose members were almost always from western states, while eastern congressmen concentrated on urban problems or foreign affairs. But with the so-called energy crisis becoming a major controversy, the Interior committee members suddenly found their own states being threatened with stripping.

Suddenly legislators like Frank Moss of Utah began paying visits to the ruined mountains of Appalachia, to look at stripping first hand. In 1972, Senator Moss spent one entire day in an army helicopter, only a few thousand feet above the scars of Tennessee and Kentucky. When he landed in Hazard, Kentucky, Senator Moss was asked (by this writer) for his reaction. The senator did not seem very disturbed by what he had seen. In fact, he said, Appalachia was actually a good place for stripping since its heavy rainfall ensured a quick revegetation.

But the senator could not get away with his Marie Antoinette "Let them eat cake" statements. Senator Moss and his committee members found themselves under growing pressure from a coalition of western and Appalachian anti-stripping zealots, small in number but persistent. Low-budget ecology lobbyists in Washington tried to match facts and efforts with the huge energy lobby. Two of the most skilled workers were Louise Dunlap of the Friends of the Earth and Peter Borelli of the Sierra Club, who hounded the committee so regularly that they seemed to know more about the stripping legislation than most of the members did. Representative Ken Hechler of West Virginia was a voice of conscience from within the Congress.

Rallies were held frequently in the mountains, with the

same small band arriving in jeeps and pickup trucks—a few teachers, a few suburbanites from Louisville or Columbus, a few clergy, the militant mountain radicals, and the few plain mountain people who had dared to ask that terrible question of whether their state and federal officials were really serving their interests. Although small in number, they kept a steady pressure on the Interior committee members.

The Interior committees did not pass a stripping bill written in 1972 but, with President Nixon busy with his Watergate problem in 1973, Congress tried again. In October of 1973, the Senate overwhelmingly passed a bill that seemed to control auger and strip mining by banning the ugly bare-rock "highwall." But at the last moment the Senate tacked on a provision that permitted the leveling of mountaintops to get the seam of coal. It wasn't clear where the "overburden" (the extra rock and dirt) would go. But Ken Hechler in Congress and Harry Caudill in Kentucky said they expected the "overburden" would be dumped on the heads of the people, as usual.

Caudill said the senators' bill was a "calamity." "They have been outwitted by the coal industry," he said.

He could have added the word "again."

Dan Sizemore has attended many of these rallies, but he has little faith in the federal legislation. Dan shivers when he hears the national leaders talking about a fuel shortage. Dan insists that the next step is for the government to move all the hillbillies onto a reservation, like they did to the Indians or to the Japanese in World War II, and gouge every last lump of coal out of the tired hills. Then, Dan persists, the government will flood the hollows to create boating and fishing lakes for the urban middle class—a New Switzerland for America—while the hillbillies live in their reservations, making quilt rugs and performing country music shows for the tourists. And when the country's leaders informed us we were running out of fuel late in 1973, Dan's vision seemed closer than ever.

* * *

Mary-Ann and Vicky notice that the sun has fallen below the opposite ridge. They expect that Margaret will be making supper. Vicky locks the rifle and carries it easily, with the barrel pointing away from us as we climb down the strip road.

"This ridge used to be as pretty as that one," she says, pointing with her free hand at the fir-treed ridge across the way. "I hear there's a big seam in that ridge, too. They'll probably strip that one, too."

Friday

19

On Friday morning Dan drives over to Milan to pick up his mail. He stops in the Phillips gas station to buy a pack of cigarettes. The station also sells pretzels and potato chips and that favorite mountain sweet, the chocolate and marshmallow concoction called moon pie.

Dan stops at the cigarette vending machine, drawn by something hanging on the wall. He raises his sights and finds himself confronted by the haunted, piercing eyes of Robert Francis Kennedy.

The station attendant sees Dan staring at the portrait.

"I've got me another picture at home," the man says. "We seen him when he came through Kentucky. Round '67 or '68, I guess. I was working in this grocery store and I heard this big crowd outside. I seen it was him. I wanted to shake his hand, but I couldn't get close enough. When I got home, I told my wife I didn't care, I was gonna hang his picture on the wall even if I didn't get to shake this hand. I've only got three pictures on my wall. Elvis, Jesus, and Bobby."

Dan doesn't mention this to the attendant, but he was over in Kentucky that day, in February of 1968, to see Robert Francis Kennedy. Only not to cheer. Dan went over there to heckle the junior senator from New York.

It was a desperate time, both for the Sizemores and the country. Pete had already gone up to Toronto; Chris had been arrested in Washington and was biding his time until spring; the Vietnam War was getting worse. The layoff and the war had turned the Sizemores into outsiders by early 1968; Dan had already decided that capitalism and profits had brought Appalachia to its knees. At that bitter time, he felt the entire outside world was conspiring to destroy him and his region.

Dan remembered the surge of hope he had felt in 1960, when handsome young John F. Kennedy had walked the tracks and visited the cabins of West Virginia. After the Mountain State had helped send Kennedy to the White House, he repaid the debt by establishing the Emergency Relief Program that provided food and medicine and temporary employment for thousands of mountain families.

But then John Kennedy had been killed, an event that is still regarded by most mountaineers as part of a larger conspiracy. At first, it seemed that Lyndon Johnson was destined to outdo even John Kennedy. With his War on Poverty, Johnson sent a flood of Office of Economic Opportunity, Neighborhood Youth Corps, and VISTA workers, Appalachian Regional Commission and Appalachian Volunteers into the region. His Medicare bill, housing programs, expanded scholarships, and food stamp programs helped keep thousands of mountain people alive.

But in the long run, Johnson was more committed to Saigon than to Greasy Creek. This commitment was to hurt the Sizemores grievously. When Pete and Chris refused to serve in the services, some young men in Bradshaw County called them cowards who had better not return home again.

Vietnam and Lyndon Johnson embittered Dan toward his country, made him look for a way to strike back. Early in 1968, before Eugene McCarthy forced Lyndon Johnson to slink out of office, Dan organized a small band of black and white youths to demonstrate against the war and in favor of Appalachian self-determination.

After making several demonstrations, Dan became known in Southwest Virginia as an emerging activist. And early in 1968 he received word that Senator Robert Kennedy was thinking of making a trip to Appalachia.

The junior senator from New York was already a critic of the war, but he was taking a Hamlet stance about whether to challenge Johnson for the presidency. As a member of the Senate Subcommittee on Employment, Manpower, and Poverty, Kennedy was holding a series of field hearings for migrant workers in California, sharecroppers in Mississippi, and Indians in New Mexico. Early in 1968, it was Appalachia's turn.

One day Dan received a message that Peter Edelman, an assistant to Kennedy, was to visit the area to gather information.

Dan gave Edelman the first-class tour of Bradshaw County. He took Edelman up on the ridge, to visit the "duck people" whose mountain dialect sounds more like quacking than English. Dan grinned slyly as Edelman tried to comprehend the language of the poor folks who huddled over their open fire, wrapped in rags and blacked by a winter of not bathing.

Later, Dan delivered a lecture to Edelman about how the Democratic party had cooperated with the coal industry in establishing a "courthouse gang" that deliberately kept the schools bad and kept other industries out. Dan recalls that

Edelman did not seem overjoyed with Dan's nonpartisan diatribe. When Edelman returned to Washington, he mapped out an itinerary that did not include a stop in Southwest Virginia. The nearest stop to the Sizemores would be Letcher County, Kentucky, far across the mountains.

Dan decided this trip was only another example of outsiders using the mountains as a dumping ground for their coal wastes or as a testing ground for their politics. Dan had had enough of being used. He decided to turn it around—and use Robert Kennedy for his own devices.

Dan and Chris discussed picketing Kennedy's little tour. They debated whether they should carry protest signs, but Chris said, "Nobody ever looks at a sign." While they were discussing an effective strategy, Chris pulled a brown-paper shopping bag over his head. Feeling inspired, Chris then cut out two eyeholes and wrote the word ANGER on the front and back of the bag. This appealed to the Sizemore sense of humor. Father and son whooped uproariously for the next hour, as they created two dozen of the hoods. Then they telephoned as many young black and white youths who were against the war. They arranged to travel to Kennedy's picnic in Kentucky. They would not say a word to anybody that day. They would not shake hands. They would merely stick together, in silent protest to generations of exploitation.

Most of Appalachia was more receptive to the young senator's plans. His slick assistants lined up a perfect tour for catching attention in the national media while testing the mood of the rural mountaineers. They also lined up every Democratic official more important than dogcatcher, to be seen with Kennedy and to build bonds for future campaigns.

The trip started with public-relations genius. Early on February 13, Kennedy's jet touched down in Lexington, where he was met by modern television crews. They would have plenty of time to follow him to the first stop, and still get the film back to their studios for the six o'clock news, a crucial item for the Kennedy staff.

The entourage sped down the beltways of the expanding horse and university town, onto the Mountain Parkway, the modern toll road that climbs from central Kentucky into the hills of the Cumberlands, ending by no coincidence in the hometown of former Governor Bert Combs, who built the parkway.

There is a point early on the Mountain Parkway where you come over a slight rise and, for the first time, notice the first gentle hills starting to swell in the distance—a moment that usually gives visitors a deep sense of anticipation.

As the hills got higher and denser, Kennedy's entourage crossed through Wolfe County, one of the poorest in the United States. But the first major appearance was scheduled down the line, in a one-room schoolhouse in a little hamlet with the catchy name of Vortex. It would provide newsmen with a terrific dateline for their first day of film and stories.

The schoolhouse was already packed when Kennedy arrived, but television crews were able to film the potbellied stove, the naked overhead light bulbs, the simple note on the blackboard: "Tuesday. Welcome Visitors."

While the residents of the little town talked about surviving on food stamps and welfare, the senator pursed his lips and muttered, over and over again, "It's not enough . . . it's not satisfactory . . . we've got to do better. . . ."

Then the television crews sped their precious film back to Lexington while the Kennedy group moved farther south, into coal-mining country.

Another major stop was scheduled at a strip mine in Knott County, where several members of the Appalachian Volunteers had promised to show Kennedy the handiwork of William Sturgill and his Kentucky Oak Mining Company.

As the two dozen cars full of staff and reporters sped toward the mountain, they were cut off by several cars from the mining company. Sturgill himself, poised and casual, was standing in the road, holding a two-way radio in his hand, a common device used by strip miners to communicate across the mountains in blasting or trucking operations.

"The senator is welcome to visit our property," Sturgill said. "We thought if he did want to see it, he would have contacted us and gone up there properly. He can see anything he wants as long as he allows my men up there."

After Sturgill's gesture, the party continued up the road. But the young Appalachian Volunteers had intended to show one side of the mountain while Sturgill's assistants were intent on another side of the mountain. A long debate was followed by an even longer traffic jam on the narrow dirt road.

By the time Kennedy got to see what the company wanted him to see, the moonlight was not bright enough for him to appreciate the fine "reclamation" job Sturgill had done. But Kennedy, often accused of ruthlessness himself, may have gained a valuable insight into the determination of a strip miner to protect his "rights."

That night Kennedy spoke at Alice Lloyd College in Pippa Passes, where he told students they could help their homeland by not leaving after graduation. He also called for negotiations to settle the Vietnam War. People remember him eating several helpings of beef stew and chatting late into the night before sleeping in the school's quarters.

The next day the tour moved to Letcher County, where Kennedy had scheduled an open hearing on conditions in the coalfields. The hardworking Peter Edelman had lined up many witnesses to appear before Kennedy and his coterie of Democratic officials.

One of the witnesses was Cliston Johnson, a picturesque mountaineer who claimed fifteen children.

"Did you ever see fifteen kids in three beds?" Johnson asked Kennedy.

"I'm moving in that direction myself," replied Kennedy, the father of ten. The crowd chuckled appreciatively, every laugh a vote.

"I just want to give you a little tip," Johnson said. "The more children you've got, you just add a little more water to the gravy."

Johnson then told about Kentucky's fabled "Happy Pappy" program, in which the men were given "training" or "jobs" in the county seat. Usually this meant pushing a broom on the courthouse steps and receiving their check from the county judge, which tended to strengthen the political hold of the judge.

"They taught us how to weigh gold and write checks," Johnson said of his training as a Happy Pappy. "Who in hell could write checks if he wanted to? And can you imagine weighing gold in these mountains?"

Kennedy seemed genuinely upset by these stories, just as he seemed upset by the poverty and deprivation around him.

"It doesn't make any sense," he said. "It doesn't make any sense to train people for jobs that don't exist."

Soon after that, Dan Sizemore and his twenty young men arrived in the crowded gymnasium. Dan had gotten permission to testify at the hearing. With his brigade of silent, hooded boys in the audience, Dan said, "This is the last time we will ever appeal to you as Democrats or Republicans. I don't know what road we will take, but it won't be a happy one."

Then Dan switched to the war. "Our area is feeding the war machine," he said. "Our boys have been taught the art of killing. They aren't genteel enough for draft-dodging."

Then, staring directly at Kennedy, Dan said that as far as he could see, most politicians consulted other politicians when they came to Appalachia. Dan said he wanted to make sure the senator wasn't taken in by this tour. Dan pointed to the local politicians sitting around Kennedy. He said these men supported the status quo in Congress and in their county seats. He challenged Kennedy to dismiss the politicians from the stage and invite poor people to sit alongside him.

Kennedy's hosts seemed to bristle at these harsh words. Men like Carl Perkins, the chairman of the House Education and Labor Committee, who always remembers to don his white socks and his bumbling drawl whenever he returns to Eastern Kentucky, were not used to being criticized within their home district.

Dan continued about the dangers of working in doghole mines, strip mining and water pollution, and people being forced from their homelands.

While Dan talked, Kennedy wrinkled his forehead and peered studiously at the paper-bag gang. When Dan had finished, Kennedy smiled and tried to coax the boys into removing their hoods. But they refused to budge. Dan, meanwhile, tried to gauge the senator's inner feelings.

"It's hard to read a feller like that," Dan said, five years later. "You could see that he wanted to care. I think the bags disturbed him greatly. It was a shocker. We were the only angry ones he saw. It set a different tone for the meeting."

Kennedy was no stranger to political anger, of course. He had walked with Cesar Chavez in California and he had talked with black and Latin leaders from the angry ghettos. Perhaps he had been misled by his staff and by the toadying manner of the local politicians into thinking all mountaineers were accepting of their lot. But Dan felt his speech set the senator in a new path.

Shortly after Dan's talk, Kennedy met with an official of Bethlehem Steel, one of the leading employers in Letcher County. The official, David A. Zegeer, division superintendent of Beth-Elkhorn Corp., tried to paint a rosy picture of Bethlehem's deep and stripping operations. Zegeer also complained about Dan's earlier criticisms of capitalism.

"The gist of what's been said is that industry is bad, that absentee ownership, whatever that is, is bad. I'd like to point out that Beth-Elkhorn has eight hundred stockholders in Kentucky," Zegeer said.

Kennedy seemed to come to life as the assertive coal executive continued to praise Beth-Elkhorn's contribution to mountain life. The senator asked how many overall stockholders Bethlehem had. Zegeer conceded it was many thousands.

"Then you would hardly say that it is a Kentucky-owned company," Kennedy said, his tone growing icy.

The two men then debated whether Bethlehem's coal mining actually gave any benefits to Eastern Kentucky. And finally Kennedy made this statement in his clipped, excited, Boston accent: "I've studied the history of Kentucky and you would clearly have to reach the conclusion that there has been absentee ownership, that outsiders have come in and exploited the great wealth of the area—with great profit going elsewhere."

This exchange was duly reported in the press the next day. Robert Kennedy raced on through Appalachia, shaking hands, attracting crowds around him, with his electric movie-star attraction. Many other people, like the service-station attendant over in Milan, put his picture up on their walls, alongside calendars of Jesus Christ, album covers by Elvis Presley, old photographs of John Fitzgerald Kennedy, and more faded pictures of Franklin Delano Roosevelt.

You hardly see any pictures of Lyndon Johnson or Richard Nixon in Appalachia.

Five years after that whirlwind tour, Dan and Margaret Sizemore sat in their living room one morning and talked about the Kennedy brothers. Time had not softened their feelings about the war or the political system. But they seemed to have a new feeling about the Kennedys.

"I was very sorry when Robert Kennedy came through here," Dan said. "It seemed like the same old shit all over again. It was a bid for votes. He was looking for misery and we had the most of it.

"I always regret the failure of our people to stand up. If Kennedy had been great enough to let the poor people take over that meeting. . . . But he was a politician. He couldn't break out of the chains."

Margaret Sizemore waited for her husband to finish. "Still, they were decent and compassionate people, Dan," she said. "Maybe they did care. Maybe they were curious. John Kennedy taught me that he wanted to help."

"Yeah," Dan said. "If Robert lived, maybe he could have opened a few doors. He could have made people try harder. He certainly would have been better than the miserable man we got. I don't know about the future. I don't see any bright spots in the future."

20

Five o'clock. That good lonely time when Dan works the office by himself, taking calls from the five sections, reading his reports, getting the feel of the shift. Dan feels isolated, elevated. It is his mine until midnight.

The phone rings, the light flashes from Ten East. The voice on the other end sounds frantic.

"Smoke, Dan, real thick. Joe and Denver went in around ten minutes ago but they ain't out yet."

Dan asks, "What did they have on?"

"Self-rescuers."

"Try to find them. I'll be right there."

Dan hangs up the phone. As quick as that, with no warning, trouble can come. Any one of a million things can go wrong in a mine, a million accidents looking for an excuse to happen.

Dan thinks to himself. He knows that all the men carry self-rescuers that fit over the nose and mouth. Enough oxygen for one hour. Yet he also knows they do not protect the eyes. In dense smoke, a man can be blinded by smoke, lose control, and just fall down, even good men like Joe Messer and Denver Hall, who are somewhere in the smoke right now, trying to find the cause.

Dan needs help. Out in the yard, Jimmie Isom, a good friend with a bad heart, is supervising a load of supplies. Dan calls Jimmie over.

"Jimmie, we've got bad smoke in Ten East. Call the expediter. Cut off all the power to Ten East. Then you take the extra jeep and bring me as many gas masks as you can find. I'm going in."

Dan hunches into his personal jeep, turns on the motor, aims it toward the shaft mouth. With any luck at all he can travel the mile and a quarter in twenty-five minutes, through the maze of tracks and switches.

The track is wide open. Dan pushes the little jeep to its limit. In twenty-two minutes he reaches the last open crosscut, where nine men are standing around. Little wisps of smoke can be seen down the tunnel.

"They're still in there," Reed Warren shouts as Dan jumps off the jeep.

Dan looks down the shaft. He has no way of knowing whether he has a raging fire or a smoldering short circuit. The only way he can tell is by exploring that shaft, where his two friends already disappeared. He knows he must go into that smoke. That is why he is paid more than his men. That is why he has been a foreman since he was a boy. The men are coal miners. He is a company man.

Dan turns around and looks at the nine men from the crew.

"Anybody wanna go with me?"

There is silence. Dan is not surprised. He has never ordered a man into smoke or fire in his life. He has asked—he has begged—but he has never ordered. Dan figures a man knows what he can do. If the man is afraid, he would only panic. There's no sense in that man volunteering.

Dan turns and walks toward the smoke. He reaches the perimeter of the curling, acrid puffs. But up ahead, emerging from the clouds, like a pair of abominable snowmen emerging from the Himalayan mists, come Joe Messer and Denver Hall, their self-rescuers still held tightly over their noses and mouths.

"It's too thick, Dan," Messer says. "We didn't see no fire but we couldn't get back to the face. It's too thick. Our eyes watered up. We never got there when the shift started today. It was already smoking."

Dan motions them back toward the crosscut. He tells them that Jimmie Isom will be arriving with gas masks. Right now, Dan wants to explore the shaft, to see if he can find where all the smoke is coming from.

He clamps the self-rescuer over his face and begins trudging inward. He walks five hundred feet, the smoke is getting thicker, Dan is aware of extreme irritation in his eyes, tears streaming down his cheeks.

Also he becomes aware of the rubber mouthpiece getting uncomfortably hot, a sure indication that the air is thick with carbon monoxide, causing the lifesaving chemicals to act inside the self-rescuer. His mouth feels as if it is being held against a hot stove. The chemicals inside the air tank, strapped on his waist, are being activated. The tank jiggles uncomfortably against his belly. But Dan is familiar with the procedure; he knows he is in no danger yet; he keeps walking toward the smoke.

Now he is a thousand feet from the nearest man. He reviews in his mind what might have caused the smoke. He knows he has a huge continuous miner, two shuttle cars, and one roof-bolting machine, all of them attached to a mobile power unit called a "power pac."

Although Dan had the power cut off immediately, one of

these machines may already have caught on fire despite the elaborate circuit breakers. And the fire could have spread to the seam of coal.

But he doesn't see any flickering shadows deep in the shaft. Dan thinks ironically, is that what Lyndon Johnson meant when he talked about the light at the end of the tunnel? The dumb bastard was walking into a shaft fire and he didn't even know it.

At eleven hundred feet, Dan's eyes start to burn. He knows he cannot go any farther without a gas mask. He must go back to the crosscut without any knowledge of where the smoke is coming from. It seems to be swirling all around him in this section.

Back at the crosscut, Dan rips off his self-rescuer and breathes reasonably fresh air. His lips are hot. His tongue has no feel to it, from the overheated mouthpiece.

Jimmie Isom has arrived with eight gas masks—large, ugly-looking devices with pigs' snouts and complex tubing and clamps and straps. But they enable a man to breathe and talk and see deep in a smoky mine.

"Let's get 'em on," Dan says.

Denver Hall and Joe Messer start strapping on their masks. None of the other men come forward to try. They stare at Dan. Dan wonders silently, why do they hang back like that? Aren't they always bragging about how a miner will protect his buddy? Maybe they would respond if somebody's life was in danger. Dan decides to ignore them. He begins strapping the gas mask on his own head.

Suddenly Jimmie Isom picks up a mask from the jeep. "Put one on me, Dan," Jimmie says.

Dan stares at his friend, with the deep-etched lines from his heart attack. Dan usually works Jimmie on the outside crew these days, afraid of working him inside. Now Jimmie is volunteering to go into the smoke. Dan doesn't know how to turn him down.

"You'd better show me," Jimmie Isom says, looking in Dan's eyes. "I don't know how to use this thing."

Dan stares at his friend. Jimmie has been working in the mines for over twenty-five years. How can it be that nobody ever showed him how to work a gas mask? There is no time to think about it now. If there is a fire, they've got to put it out. Dan tightens the gas mask around Jimmie's face. The four men pick up fire extinguishers and trudge down the smoky shaft once more. The nine crewmen wait behind, silently.

This time they travel the first thousand yards quickly. They see no flicker of fire. They keep walking. Another two hundred feet and they are in the working section known as Ten East, where the smoke seems to be coming from. But there is no fire.

Dan surveys the room. The machines are all intact. The continuous miner, the shuttle cars, the roof-bolter, the power pac—these huge looming machines that cost so much money, do the work of so many miners. The smoke did not come from them. But where?

Then he spots another machine he had not expected, a giant cutting machine twenty feet long, ten feet wide, five feet high, capable of cutting coal in this double seam. What is this machine doing in here? Dan thinks. Then he remembers. The day shift moved it in here a few weeks ago, to cut the higher seam.

He walks up close to the cutting machine and crouches down, searching for the two-inch cable that should be connected to the power pac. There, amid the mud and dust and gob, he spots a thin wispy trail of copper-colored ashes.

"Goddamn," Dan curses, audible through the gas mask. "Goddamn the son of a bitch who wired this goddamned machine."

The men stand behind him and follow his gaze. The trail of copper ashes should be directed toward the power pac. But they are not. Instead, they head toward the main power cable, the source of all energy in this room, now turned off. Where the copper ashes end, Dan stares at the power cable. Somebody has connected it to a bare feeder wire without a fuse nip—the 500-amp fuse that prevents overloading or grounding.

Every curse word Dan has ever heard comes tumbling out of him now. Sheer raging anger makes his muscles twitch. He wants to kill whoever it was that wired this goddamn machine. When the feeder wire shorted out, it burned into the cable, cutting the cable in half with a clean burn, cutting off the power automatically without anybody knowing why.

But suppose the burning cable had set fire to the cutting machine? Suppose all the grease and oil in the machine had gone up in an explosion? It might have set fire to the seam of coal. It might have ignited some invisible pocket of methane gas, just waiting for a spark. Then they might have had another Farmington, another Hyden, right here in Ten East, a bunch of poor dumb miners, cooked to a crisp.

"Goddamn, goddamn, goddamn," Dan sputters. The three men follow him as he makes a final inspection of the room, the smoke billowing around them.

Suddenly Dan hears a noise behind him. He sees Jimmie Isom grasping his gas mask with both hands.

"I can't breathe," Jimmy says. "I need air."

Jimmie tries to yank the mask off his head. He starts running down the shaft, toward the crosscut, disappearing into the smoke. Denver and Joe look at Dan. Then the three men start running after Jimmie. They make a clumping noise as they run in their heavy gear. Denver and Joe pull ahead of Dan. Dan curses his ruined lungs, curses his weary knees, curses the body that cannot run a hundred yards anymore.

He hears a commotion up ahead. He cannot run any farther. He slowly trudges through the shaft until he reaches the crosscut.

Jimmie Isom is resting on the floor of the mine, in the mud and the gob, surrounded by the nine other crew members. The miners have removed his gas mask from his head. Jimmie is gulping the fresher air of the crosscut. He is laughing and crying together, trying to talk.

"Couldn't breathe," he sputters.

Denver and Joe tell how they had found him fallen, his gas mask half off, around three hundred yards from the crosscut. Another few seconds and he might have swallowed too much smoke. Dan thinks about Jimmie's bad heart. If Jimmie is ever going to have another attack, this is the time, Dan thinks.

Exhausted from his own sudden run, Dan feels drained, sickened. As a foreman, Dan has used a gas mask dozens of times. He knows the cardinal rule of gas masks: "Faint or die, fall down from exhaustion. But whatever you do, don't take off your gas mask while you're in smoke." Yet Jimmie had never handled a gas mask before—and it almost cost him.

The men put Jimmie on the man-trip and head toward the front. Denver drives Dan's jeep for him. They reach the cold night air in the yard and Jimmie starts to revive. They take him into the bathhouse and make him rest.

Now Dan turns his attention to the mystery of the swirling smoke.

He sees Ed Williams and Larry Harding, two other bosses who have rushed up from other mines when they heard the alert. Dan tells them about the burned-out cable on the cutting machine and the illegal wiring job that probably caused it.

"Boys, I can't figure out where the goddamn smoke kept coming from," Dan says. "The cable probably burned out two hours ago, but the smoke just kept circulating."

The two bosses look at Dan.

"It was the fan," Williams says.

Dan does not understand. He stares blankly.

"The fan," Williams says, meaning the huge booster fan that circulates air throughout the section. "When we got here, we cut off the fan just a little while ago. That was recirculating the smoke over and over again. As soon as we heard you had smoke, we turned off the fan. The way that section is set up, you recirculate the bad air."

"But why?" Dan asks.

"That's the way the law sets it up," Williams answers. "We set up the air flow according to the 1969 Health and Safety Act. The inspectors keep telling us how to arrange these fans, Dan, you know that. We keep telling them it recirculates the bad air every time we've got smoke. But they tell us that's how the law is written."

Dan stares at the two officials. He is not sure whether to believe them. Whenever coal officials get together, they blame the 1969 Health and Safety Act for their troubles. Yet they often try to circumvent the law if nobody is looking. Harding himself hates the law more than anybody. Look at how he hid the bad powder inside the mine on Wednesday. Yet if something goes wrong, they blame the act.

Dan thinks to himself: Did they make some kind of shortcut when they built the air vents? Or is the law really to blame for requiring such a huge amount of air to be pumped through a mine? Dan isn't sure he knows. He isn't sure anybody knows anymore. All he knows is they almost had a tragedy inside the mine tonight. Almost a fire. Almost a suffocation of an old miner. Not to mention his own blistered tongue and lips from the self-rescuer.

In the last minutes before quitting time, Dan skulks around the office, calculating what the company has lost by tonight's little adventure. They almost lost a cutting machine, worth many thousands of dollars. They lost a thousand feet of cable, to be replaced at a cost of $1 per foot. They lost all production on Ten East, a loss of perhaps 200 tons worth at least $12 a ton. And they had to pay the men a full salary. That's at least $4,000 lost by the stockholders of the Big Ridge Company tonight, Dan thinks. Those rich bastards up in New York and Pittsburgh and Boston would love to know how we fuck up their profits.

Dan asks Williams if he is going to report the incident to the federal inspectors.

"No need to," Williams says. "Only lasted fifteen minutes. No injuries. No fire. Law doesn't require us to report this."

Williams also says he will not report the day-shift foreman who was responsible for the illegal wiring, either. That would only upset the front office over in Milan. Nobody wants the big shots snooping around, causing trouble. It would be best to forget the whole thing, Williams tells Dan.

In the parking lot, Grant Harris and Calvin Brooks open up a bottle of home-brew beer Grant's cousins have made. Dan sits in Grant's pickup truck and takes his first thirsty gulp. His lips are blistered. His tongue cannot taste the alcohol. But when it hits his stomach, it starts to burn the anger away.

Saturday

21

On Saturday morning, everybody sleeps late but Margaret. She awakens at her usual time, long before dawn, eager for her hour of peace and quiet. She sits impassively at the kitchen table, drinking her coffee and smoking. Then she shifts gears from her weekly role as college student into her Saturday morning role of house cleaner.

This morning is particularly messy. The stove quit working yesterday and Margaret hopes it is something easy, like the burned-out wire last year. Wearing her oldest dungarees and polo shirt, she grips the back of the stove and maneuvers it into the center of the kitchen floor. She takes a damp rag and swipes away a year's accumulation of dirt and grease. Then she inspects the wiring. She's in luck. The same wire has burned out again; Margaret trims down the ends and splices them. Maybe it will last for another year, she thinks as she pushes the stove back into position.

For all Dan's mechanical expertise at the mine, it is Margaret who makes the repairs at home. Dan is forever bragging about the time a repairman warned them to buy a new refrigerator, but Margaret tightened up a few things and it worked thirteen more years.

When the stove is back in place, Margaret cleans out the refrigerator of the stale salads and dried-up sandwich meats and near-empty bottles that have accumulated as the family of eight fed itself around the clock for the last week. Then Margaret inspects the pantry and makes out her shopping list.

Saturday is shopping day in the coal country, the only day when husbands are home and markets are open and paychecks are still intact.

When the Sizemores had no car, they used to shop by phone. But the only store that would deliver was the Big Ridge store in Milan, at prices that were often 5 or 10 percent higher than in Morgantown.

Margaret frowns as she makes out her list. It is almost impossible to shop for this family: David is crazy about meat. Vicky won't touch any meat and exists mainly on peanut butter and banana sandwiches. The boys insist on snacks. Dan needs things for lunch and he loves to have a snack when he comes home late at night. Plus, it's hard to make everything fit into the refrigerator.

The sky is bright now. Margaret has been awake for three hours. David is the first one down, just before nine o'clock, chattering how Dan promised to take him to the mine today. David never forgets anything promised to him. Edward and Gene drift down later. Then the girls. Bobby, of course, will sleep indefinitely.

Dan pads into the kitchen at ten o'clock, poking at his blistered lips, telling the story of the fearful smoke scare the night before. Margaret serves him eggs and french toast and coffee.

Dan sits at the table in his bathrobe, feeling delighted to be telling his story to a full audience for a change, happy to see Margaret home.

"Are you working today?" Margaret interrupts his tale.

"Why, of course," he replies. "They didn't get their coal out last night. They'll be sure to want a crew cleaning up the section today."

"You promised David a trip to the mine," Margaret says. "He's going to hold you to it."

Dan nods. As a foreman, he is on call whenever Big Ridge wants him. Theoretically, the company expects him to work every other Saturday. In practice, he must visit the mine almost every Saturday, if only to get things started and supervise the cleanup. The miners, of course, are working on overtime whenever they work on Saturday, while foremen get nothing extra. From time to time, there is talk of organizing a foreman's union. Dan recalls how he was trying to avoid just such an organizer the time he met Margaret.

"I thought we'd go shopping this morning," Margaret says, when everybody has eaten.

They pile into the car, the two girls, the two little boys, Dan and Margaret. Dan lets Vicky drive. She has been eager to drive since she was around fourteen. She drives slowly and carefully over Milan Mountain, Dan fussing at her in the front seat, then pulls into Milan to pick up the mail. The old company town is packed on Saturday morning, pickup trucks jamming the narrow street, shiny new cars gunning their engines as they wait for a parking space. The noise is intensified at the bottom of the valley as the tiny wooden bridge rumbles when a car crosses it. Miners in clean sport shirts stand on the sidewalk and trade gossip from their various mines. Dan stands in front of the Big Ridge Company Store, with its bright assortment of lamps in the window.

The days of mandatory trades are long since gone. Nobody gets paid in scrip anymore and nobody owes his soul to the company store, unless he is a fool. But the company is not doing any favors for the residents of Milan. One glance in another window reveals the price of disposable diapers, a product I am more than familiar with. A brand that normally sells in supermarkets in Louisville or New York at between eighty-nine and ninety-nine cents per dozen sells for $1.29 in the Big Ridge store. Which is why the Sizemores don't shop here.

Down the sidewalk is Calvin Brooks, all dressed up in green slacks, red checked shirt, and a white golf cap over his graying black hair. Calvin carries himself with the poise of a retired athlete, who has lived in the spotlight all his life. Dan purrs happily as he shakes hands with the only black man at his mine. There are few people in the world he loves as much as Calvin. Dan often visits Calvin at his home on the "colored hollow" that stretches north from Milan, identical wooden cabins on both sides of the narrow road. The stripping above Calvin's hollow has caused most of the yards to be washed out, with debris scattered in the grassless yards. But inside Calvin's home everything is sparkling, with religious pictures and bright objects displayed in every corner.

Dan recalls the last time he visited Calvin's home. Calvin was drinking some quality moonshine while Mrs. Brooks was reciting the Bible at him, trying to change his ways.

"Dan, help me with this woman, will you?" Calvin had asked.

"Calvin, I can see the situation clearly and I must tell you—I'm on your wife's side," Dan had said, laughing, as he beat his retreat.

But Calvin bears Dan no ill will for bailing out that day. In fact, he informs Dan that several carloads of his relatives have come down from Detroit and will be holding a family reunion this afternoon.

"Dan, you bring your whole family over this afternoon," Calvin says. "I mean it now. You know you're welcome."

"Calvin, I would love to drop in but I'm afraid I've got to

go over to the mine today," Dan says. "Reckon you won't be working today?"

Calvin says he doesn't need the overtime that bad. When he shows up, Calvin is the best loader-operator at the Big Ridge mine. But he often stays away Fridays and Mondays, too. He can get by on three days of work. More power to him, Dan thinks.

The Sizemores drive out of Milan, sharing the mail. Today is a rich crop. Since the Sizemores became estranged from most of the community, in those past few dangerous years, the mail has become increasingly important to them.

They all thrill when they spot a letter from Pete and Chris in Toronto. And they welcome mail from a woman writer friend in New York or a young couple who organize for a steel union in Georgia. They get frequent mail from activist friends at Virginia Tech who inform them of the latest strip-mining or black lung protest. The mail helps to narrow the distances across the long ridges between them and their friends.

The ride into Morgantown takes around fifteen minutes, just long enough to run out of coal country. The improvement is apparent immediately. No more scars on the mountains. No more film of dust on everything. The roads are not as badly worn from the overloaded coal trucks. No rotting tipples by the side of the road. Even the houses seem more substantial. Morgantown is a pleasant middle-class town of a few thousand. It makes its living at the fringes of the coal industry while managing to avoid looking like a coal town.

The town seems almost festive on this bright Saturday morning. They park the car in the lot of the Piggly Wiggly store. In the pickup truck in the next parking space, two hogs are penned in the back by some wooden slats.

While Margaret and Mary-Ann shop for food, Vicky takes the two boys to a toy store for their Saturday morning treat—comic books and maybe a small toy or a record, followed by a soda or candy. The mood is cheery, like that of most small towns in America on Saturday morning. The gloom of the coal industry is briefly dispelled.

Dan stands on the sidewalk and waits. He hates shopping. The only reason he goes is to help Margaret move the groceries and to supervise Vicky's driving. But it drives Dan absolutely out of his mind to shop with Margaret in the Piggly Wiggly, watching her price each item. He knows it is important to price the items. But he cannot bear the strain. He swears his hips ache more in fifteen minutes of standing around the Piggly Wiggly than after a full shift over at the mine.

He feels much better about standing on the corner and watching the Saturday morning parade of pickup trucks, people loading up on fertilizer and Purina dog food, tow-headed little boys in T-shirts with toy guns in their belts, holding on to their fathers' hands. Dan looks across to Bensons' department store, remembering the last time he went inside. All he wanted was a sport shirt, but the saleswoman talked him into buying an album of religious organ music by her son, who was studying to be a Baptist preacher. Dan felt so foolish about buying the record that he felt he had to get his money's worth by playing it. He discovered that the boy had a good feeling for revival-type music.

The streets of Morgantown are having their own revival this morning. Prosperity has arrived, in the form of paychecks, welfare checks, disability checks, food stamps, all on the first of the month. Dan feels inspired, expansive, about the waves of people. He decides to buy a bottle of wine for Sunday dinner.

We go into the State Liquor Store, run by the state of Virginia. The bottles are stacked out of sight in the back. There is no advertising. One sour-looking clerk sits behind the plain wooden counter. He stares at us. We ask if he has any Inglenook white wine. The man points at a large sheet of paper taped to the wall.

"Everything's listed over there," the clerk says.

We glance at the list of wines. A few reds, a few whites, half of them crossed out. No Inglenook. Very little else. We settled for a bottle of cold duck. The man hands it over the counter in a brown paper bag, looking as if he were selling dirty French postcards, against his will.

"That man's got the perfect disposition to teach in our county school system," Dan drawls after we leave the store.

Outside, the town is growing more crowded as noon approaches. The taxi stand is swarming with people, welfare clients who must rent a taxi to shop when their checks come in. The taxi business has boomed in Appalachia since the government ruled that welfare money could be allocated for travel to shopping or medical care. The taxi drivers now charge eight or ten dollars for a few short miles, knowing the government is paying for it. Some fleets have jumped from a few sputtering cars into dozens of new machines. Free enterprise is not dead, it has just taken a new form—symbiosis. Follow the government and start nibbling.

Dan wanders back toward the Piggly Wiggly. Margaret and Mary-Ann are pushing large baskets loaded with milk cartons, soda cases, canned vegetables, packaged sandwich meat, some fresh vegetables and fruit, bags of bread and snack food and cookies.

"I wasn't going to buy any meat," Margaret says, half-apologetically. "But the pork chops looked real good."

The Sizemores drive to the next block and find Vicky walking with Gene and Edward. Several young men have stopped to talk to Vicky. They look disappointed when her parents pull up. Vicky and Edward and Gene get into the car for the ride home. Dan runs his finger over his face as he drives with one hand. Margaret talks about the high price of food, her voice growing monotone from exhaustion, taking on her trembling, half-questioning tone.

Back home, they all help carry the dozen packages up the

steep stairs into the house. Margaret and Mary-Ann clean up the kitchen and make lunch. Dan opens a beer and rocks on the porch swing. He holds the cold beer bottle against his burned lips and closes his eyes. He has an hour before he goes to work.

22

Every few minutes, Dan Sizemore paces nervously into the parking lot, scanning the darkness for the headlights that will tell him David, Bobby, and Vicky are arriving. He expected them half an hour ago. He is apprehensive whenever anybody tries to drive in the darkness on the wretched strip roads.

The family rarely visits the mine, although the miners have grown used to seeing Dan's long-haired friends from the cities and the rumpled mountain activists and the new breed of women. (One trio of liberation types tried to give the miners a rap about how women and miners were both part of the same oppressed class; it didn't go over.)

Margaret hasn't been over in five years. The last time she visited, old Jimmie Isom, keeping a straight face, tried to console Margaret on her being blind. She must be blind, the poor woman, Jimmie had drawled. Otherwise, why would such a pretty woman be married to such an ugly man as Dan Sizemore?

It has been even longer since Mary-Ann saw Dan in his dirty mining outfit and burst into tears. She has avoided the mine ever since.

But David and Bobby and Vicky enjoy talking about the mine. David pastes pictures of coal cars and freight trains on his bedroom wall. Bobby once said he wouldn't mind working in the mines, but Dan refused to consider it. And Vicky says she would like to be a woman miner. ("She doesn't even do her work at home," Margaret laughs.)

Things are especially quiet on Saturday night. No bosses, no federal inspectors, not even a full production crew tonight. The deserted office and mine yard take on a magical, fairyland setting. An amusement park that can blow everybody up, Dan thinks.

Just before six o'clock, Dan sees the headlights working their way up the mine road. He waits out in the cool night air in the parking lot.

"I was worried you'd gotten lost on those strip roads," he says.

"We had to stop for gas," Vicky says.

Peeking first to make sure nobody is inside, Dan escorts the children into the bathhouse. He points out the shower room, where the men would be splashing Joy soap all over each other. He lowers a clothing basket from a pulley and tells them, for the tenth time, how he doctored up the boots of miserable little Estill Dean.

Then he takes them into the office, where they inspect the mine map, with its thousands of pencil marks for pillars and shafts and air vents.

"That's where we had our trouble last night," Dan says, pointing out Ten East, deep in the bowels of the mine.

The children are mute in the fairyland. The office seems warm and bright after the dismal darkness of the strip roads.

They amble around the office, feeling like children playing hooky. They know that Dan would only dare bring them into his office on a nonproduction Saturday. In the car on the way over, they have been debating whether Dan will actually be able to slip them into the mine. They do not even dare to ask him the question. They will wait to see how things work out.

Good thing they do not ask just then. The door is opened and a wrinkled foreman named Buford Dark walks in. Buford's red-lined drinker's face breaks into a practiced smile when he sees the three children.

"Well, there, Dan, you fixing to recruit these three for my section?" Buford asks heartily.

The three children smile carefully, more suspicious than shy. Dan makes the introductions as Buford sits down at the center table and opens his lunch pail. But before he takes a single bite, he jumps up and begins arranging chairs around the table.

"Here, miss," he says to Vicky. "You sit down here with me and have some of my supper."

The children filter around the table. Buford reaches into his pail and finds a pack of six doughnuts. He passes them around insistently. The children look at Dan.

"Buford doesn't usually share his supper with man or beast," Dan says. "You'd better take him up on it. Else he'll wolf down all six doughnuts like he usually does."

They each take a doughnut. Buford hunches over a sandwich, eating in quick bites. When the sandwich is done, Buford says, "Ask your daddy to take you into the mine. Be quite a sight if you've never been in there before."

Dan sizes Buford up. He wonders if Buford would whisper to the big bosses if he discovered that Dan really did take his children into the mine. Dan suspects that Buford resents him ever since he was busted from regular foreman to just an extra foreman.

"I think we'll just walk around the yards and I'll point out some spare equipment to them," Dan drawls. "Thanks for sharing your supper, Buford. Knowing how you love to eat, it was a supreme sacrifice."

"Anytime," Buford says. He bows as Vicky and the two boys follow Dan out the door. "Nice to meet you, miss."

In the darkness they pick their way over tracks and greasy puddles and discarded timber. This yard is getting worse all the time, Dan thinks. Nobody cleans up anymore. They pass the powder tank, where Estill Dean and Grose are loading their bubble duster, the car that carries the rock dust into the mine. Trim little Dean salutes the family with a crisp motion of his hand, the same quick salute that can flick a water bag fifty feet.

The powder is funneled from an overhead tank into the bubble duster. When the car is full, Dean and Grose navigate it into the mine mouth. They will deliver supplies deep in the mine for the next few hours. Dan knows the only other crew is out cleaning up Ten East. The yard should be quiet for the next few hours.

"Well, David, are you ready for the ride in the jeep?" Dan asks.

David is subdued, the way he gets when confronted by sudden change. He hasn't said a word since the car pulled into the parking lot. He is stooped lower than usual, his lips working nervously.

"Come on, David, let's take a ride in the jeep."

Dan slips behind the controls of the small jeep. He pats the seat next to him. David sits down hesitantly. Dan sets the jeep in motion, a few sparks flying from the conducting arm. The jeep lurches forward slowly, curving into the concrete mouth of the mine. The jeep sounds clackety-clack as it crawls into the mine. The headlight reflects off the white walls. Then the concrete gives out, the light is diminished, flickering eerily on the rough walls. Dan pilots the jeep a few more feet, then stops. He looks at David. The boy barely looks from side to side. Dan throws the jeep into reverse and backs out into the yard again.

"Well, David, now you had your ride inside the mine," Dan says.

"I rode the train, Daddy," David says slowly. "I rode the train in the mine." He begins to smile.

Dan hops off the jeep. He starts walking on the tracks that lead to the mine, motioning for the children to follow him. Bobby turns to Vicky. It is going to happen, they nod to each other. Dan's cap light illuminates the shaft, while Bobby picks up a hand lantern. They all walk a hundred yards of concrete and then they are surrounded by rock and timbers.

As they walk awkwardly over the railroad ties and the puddles, Dan tells them the story of the first man to die in this mine. The children can visualize the bunch of miners hunched over their stricken companion, their lamps casting a glow on his thrashing body.

"They say that a man's soul stays in the mine," Dan says. "We had a black man named Willie get squashed to death one time. For months afterward, you could hear these horrible moans inside. Sure, you could argue that it was the wind or maybe some animal that had fallen through a cave or something. But the miners were positive it was poor Willie cursing the Big Ridge Coal Company for causing him to be crushed."

"Do the men really believe that?" Vicky asks. They have discussed this subject a thousand times, late at night in the living room. But never in the eerie shadows of the mine. Dan's answer takes on a deeper meaning as they walk deeper into the mountain.

"I believe most miners believe it," Dan says. "Too many things happen that can't be explained. It's funny, though. Most of these men are religious, too. They believe God will take care of their souls. But when they hear that dreadful rumbling from deep in the mine, when a roof falls or a rib buckles, they don't say it's God. They say it's Big John. They say Big John is rumbling down there."

Dan walks a few feet more, to the first working crosscut. On a production day there would be a full shift of dirty-faced miners scurrying around here, running from one section to another. Dan points out the conveyor belts and the air shafts and the roof bolts plugged into the ceiling. He points out the power boxes, with their complex cables. Then he turns around.

"I believe we'd best get out now," he says. "This is far enough."

Although he has not discussed it with the children, they can sense he is apprehensive about being seen with his children in the mine. Dan is controversial enough to the big bosses down in Milan. They have heard about his speeches to activist groups on the evils of the coal industry. He doesn't want to give them any specific reason for letting him go.

Vicky takes a last look over her shoulder. David seems relieved as they emerge into the yard again. It has started to rain. They rush into the lamp room.

In the lamp room, two belt repairmen are eating their supper—big Elmer Spicer, a huge bachelor who works every overtime Saturday and gives his money to his sister, and Lew Begley, the half-Cherokee who plays a great guitar with his three remaining fingers on his right hand. The two belt men have their feet up on opposing benches, eating their supper leisurely.

"Got some new miners, eh, Dan?" Elmer asks. He takes a baloney sandwich out of the wrapper and offers it around. The children nod, no thank you.

"Well, did you visit the mines?" Lew Begley asks in his big, booming voice.

"Oh, we just kind of walked around the yard," Dan says.

"We could run 'em on back in our jeep," Spicer offers.

"Thanks anyway," Dan says. "But they'd best be moving along."

Dan walks the children out to the parking lot. His mood is hushed and conspiratorial.

"It was real good fun," Vicky says.

"Well, I think you saw everything," Dan says. "Tell your mother I'll be home early, around ten or eleven. Tell Edward and Gene we'll tell us some ghost stories when I get home."

The car rolls away into the black strip roads. Dan goes back into the warm yellow lights of his vacant office.

Our car rumbles over the unpaved roads. The rain pelts against the window. We are all silent, full of the magic of the secret trip inside the mine.

After we reach the top of the ridge, we switch on the car radio, to keep in touch with civilization. It is Saturday night and the local station is carrying the Grand Ole Opry, direct from Nashville, Tennessee. It is one of the grandest traditions of rural America, the foot-tapping, guitar-picking country music coming from the old auditorium in the skid row of Nashville.

Country music gone national now, all over the national television, with Johnny Cash and Loretta Lynn popular in California and Wisconsin and even Up East. But the roots of country music are up in these weary old hills. The Carters and Stanleys are only two of the most famous country music families to come from the churches and music shows around Bradshaw County. June Carter, Johnny Cash's wife, told me once how her family gave a show in Milan when she was so tiny she slept in a guitar case.

The Sizemore children like rock and folk music more than country, which they identify with the local power system they resent. But they listen to the Opry radio show tonight, drawn by the strong male voice of one Tom T. Hall.

Tom T. Hall is one of the new breed of country singer. He comes from over in Eastern Kentucky and he makes the same good twanging sound of Hank Williams or Marty Robbins. But Tom T. Hall is also a modern storyteller. He sings tonight about sitting in a bar in Florida, waiting for his jet plane. He sings about the time he nearly starved in a motel in Roanoke, waiting for pay day. He sings about the mine that blew up in Hyden, Kentucky.

Over the radio, you can hear the audience applauding and letting loose with a few scattered rebel yells. It sounds raucous and full of life, the way country music should be. But I've been to the Saturday-night Opry Show; most of the audience is so exhausted from driving a few hundred miles for the show that they sit placidly, fanning themselves in the unventilated old hall. Or they are in such awe of being in the presence of Roy Acuff or Bill Monroe up on the stage that they act as if they were in church. (The old Ryman Auditorium was a church, in fact, before it was converted into the Opry Hall.) Thus much of the excitement of the Saturday-night radio show is a product of the sound engineer. Anyway, they are about to move the show out of downtown, to a glittering new amusement park on the edge of Nashville. For the folks sitting by the radio, every Saturday night, the show will probably never change.

Tom T. Hall sings a song about someone named Clayton Delaney dying.

Whiskey and early death. Everything is normal in the mountains on Saturday night.

Back on the main road, we discuss stopping for a bite to eat. Bobby and Vicky debate which places are safe.

"Fast food and fast bullets," Vicky says about one spot.

"Bucket of blood," Bobby says about another.

We stop at a roadside drive-in where the teen-agers congregate. In this era of ecology, the parking lot is littered an inch deep with crushed soda cartons and paper bags. Inside, we sit in a booth speckled with ketchup stains. The waitress takes our order, stony-faced. Her hair is all done up, for a date later in the night. Something about each of us seems to turn her off: my beard, Bobby's long hair, Vicky's direct stare, David's dreamy smile. The waitress jabs her pen at her order pad. Everything fried in grease.

Too bad it is not Sunday dinner, the only time you can be confident of buying good food in the mountains, when many country restaurants put out a super assortment of country ham, fried chicken, fresh vegetables, sweet potatoes, and corn bread, just as church is breaking. But the rest of the time the food is frozen or canned. (After developing a massive love affair for greens, which are served in mountain restaurants, I learned that in most cases they come right from a can.)

The waitress slams the food down on the table. My little square fried-fish sandwich is so hot from the grease that it burns my tongue.

Back on the road again, the mist moves down the hill, cutting away our depth perception, making the road seem like a cardboard cutout, with nothing beyond it but the mist. The road curves through unseen valleys, hugging the bends in the creek, our car a submarine inspecting a murky sea.

We twist the radio dial for a hint of rock music, hoping for James Taylor or Janis Joplin. The Opry show is off the air. Everything else is either mediocre country or top-forty pop music. Did Stravinsky ever really exist? Did the Beatles really die to save our souls? It is like living in a time warp. We flick the radio off and cruise home in the mist.

Back home, Margaret is studying for her exams next week. David tells her about his ride in the jeep, chattering now that he is back in familiar surroundings. Vicky turns on the stereo and we sit around in the living room, everybody else smoking.

Mary-Ann, Vicky, and Bobby spend most of their nights at home. They say it is too dangerous on Saturday night to go driving with their friends because some of the young men in the valley stake out a long straightaway as their private drag strip. When they get the signal from the top of the hill that nothing is coming, the two racers will peel out from a standing position, leaving behind a stench of burning rubber and gaseous exhaust. Woe to the driver who blunders out of a side hollow road while the two dragsters are accelerating toward 100 m.p.h. The Sunday and Monday papers almost always carry the name of at least one Southwest Virginia boy who went out of control at the wheel late on Saturday night, sometimes taking a few strangers with him.

Once in a while Vicky and Mary-Ann will go for a drive with a few boys they like. They may go to the nearest movie in Morgantown, an hour away, or to one of the littered drive-in restaurants.

Sometimes Vicky and Mary-Ann decline invitations to go out, particularly when their parents' friends are visiting from Atlanta or Blacksburg or Washington. They enjoy sitting around and hearing news from "the outside."

Vicky is often critical of the children she grew up with. She says that in her four years of high school she met one boy she thought really had something on the ball. He's away at the University of Virginia now.

"There were three teachers that were pretty good," she says. "Most of them don't know anything. They're young and from around here. They just don't make any sense when they talk.

"When the campers from up north stayed here a year ago, they talked about their schools up north," Vicky continues. "They talked about making films or having open classrooms. It sounded nice. I think I'd like to learn under those circumstances."

Her own Bradshaw County High has been threatened with loss of official credits by the state education association because it does not provide enough college courses, Vicky says. The only way the school made it this year was by counting vocational courses. Vicky has graduated just in time. Next year she'll join her mother and sister at college.

"Only around eight kids in my class of seventy are going to college," Vicky says. "My girl friend is good in artwork. She could get into an art school, I'm sure. But she says she's had enough of school. She wants to go to work in a dress factory right away. I think school ruined her head. Most of the boys just go in the service. We can't wait to get out of school."

The children discuss a friend of theirs who has been arrested four times for possession of marijuana, which has always grown plentifully on the slopes of Appalachia. In World War II, "hemp" was a valuable cash crop, but that was before young people smoked it. The county court has not sentenced their friend yet. The Sizemores do not think the judges are giving penalties for grass, even in Bradshaw County.

At 11:30, Dan trudges in the front door, a foul mood all over his face. One of the supply cars broke down, he says. It was strictly an accident but they couldn't get Ten East cleaned out tonight. He thinks the car derailed because it was not properly maintained.

Dan seems angrier tonight than after he burned his lips last night. He goes to the refrigerator and opens a can of cold beer. The children get the message. There will be no telling of ghost stories tonight. They filter up to their rooms and turn on their music real loud. Dan sits in the living room and drinks the beer until Margaret has finished studying and joins him. They sit up talking until the house is quiet.

Sunday

23

At ten o'clock the bells start ringing at the Holiness Church of Jesus, down in the camp.

Rocking on the porch, Dan and Margaret can see the cars pulling alongside the white wood-frame church. The women are wearing their best dresses; the men are wearing suits with their shirt collars spread outside the suit jacket. The cars are mostly freshly washed, giving a feeling of rebirth in the dusty coal valley.

Since they stopped going to Catholic church nearly a decade ago, Dan and Margaret have not been near a church except for an occasional strip-mining rally. Dan is openly atheistic; Margaret says she lost her faith. They don't talk much about it anymore.

The Sizemores are not unusual in staying away from church. Although people associate Appalachia with the "Bible

Belt," actually the "old-time religion" is just one strain of mountain life. Preachers and schoolteachers were late in coming to the mountains, usually decades after the settlers had built up their mountain ways. Many of the mountaineers carried resentments against the established religions of the East Coast or the other side of the Atlantic. When religions came, they attracted a fervent minority of true believers. But nowhere in this entire schizophrenic nation are the sacred and the profane more closely intertwined than in Appalachia. Families that devote half their waking lives to church activities live next door to families that could not care less.

Dan usually has a few parttime preachers working at the mine. He tries to keep them in different sections, so they won't argue religion when they should be bolting the roof.

Once he heard Preacher Stamper telling Preacher Davis, "I always had a lot of respect for you—until you dynamited Preacher Haskins's shithouse."

Many miners turn to preaching when an injury ends their mining careers. It is quite common to see a preacher gesturing with a hand that is missing a finger or two, or hobbling vigorously back and forth in front of the congregation.

Congregations are small in the churches of Appalachia. No more than fifty persons have entered the Holiness Church of Jesus this morning. But on Easter Sunday or Decoration Day, when all the relatives come back from the factories up north, the church cannot hold all of them. Services are held on the cemeteries up on the ridges on those holy mountain Sundays, with families holding reunions to honor their ancestors now nestled in the arms of the Lord.

The churches preach the religion of the Judgment Day, of righteous anger of the Lord, of holy miracles, of being born again, of getting the spirit, of speaking in tongues. There are even a couple of serpent-handling churches in isolated corners of the Southern mountains, where a few anointed souls cradle the deadly copperheads and rattlers, chanting their faith in the

Lord and keeping both eyes open for the unpredictable.

The preachers preach in chanting fashion, accentuating their message with grunts and groans of true emotion:

"And only the–uh!–true believer can–uh!–climb to the top of the Lord's holy–uh!–mountains. The Lord said–uh!–whosoever shall believe in me–uh!–shall be saved from hell-fire. Do you–uh!–want to be saved?"

And the congregation shouts back that they do, indeed, want to be saved.

For those stuck back in the deepest hollows, or bedridden, or otherwise unable to make it to church, the small community radio stations broadcast the services live on Sunday morning, sandwiched between gospel music.

Those are the backwoods, mountain churches. In the larger towns, there are more sedate Methodist and Presbyterian churches for the teachers and lawyers and undertakers in the middle class. In areas where Eastern European immigrants settled, there will be more Catholic churches. But when mountain people think of Sunday, they accept the hard-breathing preaching as their roots, even if they haven't been to church in years.

Sunday is also a day for visiting family. The fresh-washed cars, with a new film of dust on them, will be parked in front of relatives' houses all day, with the smell of fresh chicken and corn bread wafting through the valley.

The Sizemores do not expect visitors today. At least, not family. Margaret's mother has not come up from North Carolina in several years, although she continues to write and telephone her love and support.

Dan's brothers and sisters have not seen him since Pete and Chris went away in those violent years at the end of the sixties. Thus, when it looked as if the community really meant to kill them, Dan and Margaret had already been cut off from church and family; the persecution seemed like just the inevitable next step.

24

Just as violence came to Watts and Chicago and Kent State, so it came up the path to the Sizemore house.

The family had become more and more isolated for a variety of reasons—Pete and Chris going over the border, their break with the Church, their involvement in the citizens' organization, and the increasing parade of out-of-county license plates that poured up the narrow dirt road to visit the Sizemores. Gradually the people in the county began to identify the Sizemores as radicals, people to be watched.

But it was one special visitor who touched off the violent years. Soon after the Sizemores came back from spending Christmas of 1969 in Toronto, they got a call from Chris asking if a friend of his could live with the family for a few weeks.

The Sizemores knew Charlie York as a black boy with a

reputation for fighting with whites when he was in high school. Charlie was not welcome in his own black hollow because his wild afro haircut was not yet in fashion among mountain blacks and because it focused the attention of white thugs. Also, Charlie was rumored to be on drugs. But the Sizemores remembered that Charlie and Chris had always been friendly in high school. As a favor to Chris, they agreed to let Charlie stay with them for a while.

Charlie came to the house early in January. He said he was awaiting some back pay that was being mailed down from Detroit. As soon as he got that, he was heading to the West Coast. Charlie appeared nervous. He said he had broken his drug addiction while back in Detroit because he had grown frightened about his future if he stayed on heroin.

The Sizemores were touched by Charlie's story. They moved David and Bobby in with the two little boys and let Charlie have a bedroom to himself. After two weeks, Charlie seemed more calm. He began spending more time downstairs, talking with everybody about his plans for the West Coast.

Charlie had not ventured outside the home since coming over. But late in January the Sizemores were pleased when he offered to accompany the Sizemore children to a basketball game at the high school. They would all take a special bus the school ran through the valley before and after the game.

The trip to the school was uneventful. But at halftime of the game, the principal and a deputy sheriff, Clyde Turner, summoned Charlie outside the gym. The deputy warned Charlie to get out of the Sizemore house and get out of Bradshaw County, because whites didn't like niggers living in white areas.

("I'm sure that sex was the basis for it," Dan has recalled. "The Southern mind is sick about sex. They assume that any black man is just interested in whites for their women. See, the Southern people still think their daughters are Southern belles. Their mothers start giving them permanents when they're eleven years old. The fathers look to sell their daughters off to the highest bidder. It's no wonder they're screwed up about sex.

("Sometimes I think that's the exact reason why Southern white women do get attracted to black men—because blacks are more natural about sex than sick white men. They're more tender and compassionate. They don't push sex as a means to something else. But, of course, that had nothing to do with Charlie. He was like one of the family.")

After this warning by the deputy, Charlie stayed for the second half of the game. On the way home, the bus driver ignored the state law and switched off all the lights inside the bus. Apparently, this was a signal for the boys to work Charlie over. But Charlie stood up and took off his jacket and invited somebody to start something. The white boys backed off.

When the children got home, they told Margaret what had happened. Margaret saw that it was not yet eleven o'clock. It would be at least two hours before Dan would get home. Margaret reasoned that it was not unlikely that the men of Bradshaw County would take matters into their own hands.

Trying to sound calm, Margaret sent all the children upstairs. Then she pondered her own move. Her first thought was that she could avoid all trouble to her family by asking Charlie to slip out of the house immediately. But then she felt so sick and ashamed of herself, for thinking of turning him out, that she vowed to protect his life with her own, as she would for any of her children. Margaret would later refer to this moment as "the moment I joined the human race."

Quietly, so as not to disturb anybody else, Margaret tiptoed upstairs and summoned Vicky. Normally Margaret would talk things over with Mary-Ann, because the two of them could communicate almost without saying a word. But Vicky, who was sixteen at the time, was the only member of the family besides Dan who knew how to fire the shotgun.

Vicky laid the shotgun on the table and showed Margaret

how to load it. Then she demonstrated how to aim the shotgun and how to squeeze the trigger. For the first time in her life, Margaret picked up a loaded weapon. She walked into the darkened living room and sat by the window that faced downhill, toward the grove. She vowed she would shoot anybody who tried to storm the house.

Two hours went by. Then Margaret saw Dan's flashlight, pointing his way up the path. When Dan came into the house, Margaret explained what the deputy had told Charlie. Dan stormed to the telephone and called Clyde Turner on the telephone. The deputy denied making any threats to Charlie but insisted he had merely been checking Charlie's draft card to see if he were a draft-dodger. Dan knew that Clyde was making a clear reference to Pete and Chris. Boiling mad, Dan told the deputy he would pursue the matter in the morning. Then he slammed down the phone.

Dan and Margaret sat up all night in the living room, waiting for the gang that never came. But Margaret's instincts had been correct. Two years later they met a stranger who said he had been with a group of twenty men that night who had planned to run Charlie out of town. But the hour had gotten late and the whiskey had overdone its magic and the enthusiasm petered out. The stranger said, two years later, that fully half the men in that party had never met the Sizemores but knew them only by reputation.

The next day was Saturday. Dan was supposed to run a rock-dusting crew that day. So he decided to postpone his argument with the deputy for another day.

In those days, Dan took a ride to the mine from Joe Brown and Webb Harrington, two members of his Saturday crew. Without a second thought, Dan walked down the grove and got in Joe's car, as usual. As they began to drive to the mine, Dan noticed two rifles stashed behind the seat. Joe explained them by saying he might do some rabbit-shooting up on the ridge on Sunday.

Dan also noticed two whole fifths of moonshine exposed in the back seat. Joe said they had just purchased the bottles. But Dan thought that Joe was usually too broke to afford a Coca-Cola.

When the car turned on the ridge to begin the long, lonely descent toward the mine, Joe stopped mysteriously at a little grocery store and returned with two bottles of pop—one for him, one for Webb, but none for Dan. Dan began to get a feeling that something was wrong.

A little farther down the road, they came to a break in the woods. Dan saw a car parked in the clearing, but the winter sun was shining at such a sharp angle that he could not recognize it.

"There's Clyde," Joe said casually, stopping the car.

As Joe turned off the motor, the deputy walked around to the passenger side of the car and began shouting at Dan. Still slow in realizing what was happening, Dan rolled down the window to hear Clyde better. The deputy smashed him across the temple with the barrel of his pistol.

Before Dan could react, Clyde jerked the door open and started slugging Dan with the pistol, over and over again. Dan crouched down in the car seat, trying to cover his head with his arms. He could hear Clyde cursing and screaming how he hated niggers and draft-dodgers and Communists.

Then Clyde stepped back and shouted, "Shove him out, I'm gonna get him right now!"

Dan felt himself falling into the dusty road. He tried to roll toward the edge of the incline, hoping to roll away, so Clyde could not shoot him.

Just then, they all heard a car kicking up the rocks as it climbed the hill. The car was a surprise because it was still too early for the skeleton Saturday day shift to be letting out.

Somebody grabbed Dan and shoved him against the side of the car, trying to hide him from view. Then Dan heard Joe and Webb telling Clyde to quit off. They put Dan back in the car, with Dan struggling to keep his head above the window level, so passengers in the other car could see his bloody face, in case he needed witnesses later. But Dan never knew if the other driver saw him; the car never stopped.

When the car had passed, Clyde Turner leaned into the car and said, "If you ever tell anybody about this, I'll get you."

Joe continued the drive down to the mine. Both Webb and Joe insisted they had not known Clyde was planning to whip Dan and said they never would have stopped the car if they had. But Dan knew in his heart it was bullshit. The men had planned for Dan to be shot up on the ridge, with moonshine whiskey on him, so Clyde could say Dan Sizemore had been found drunk coming to work and was shot when he resisted arrest. It was a trick that had been used before in Bradshaw County.

When the car reached the mine, a few men asked Dan how he had bloodied his face. Fearing instant repercussions, Dan insisted he had been in a minor car wreck but that he felt fine now. Then he wobbled into the bathhouse to wash the blood off his face, while the two men got dressed to work. Somebody later told Dan that young Webb Harrington had broken down and cried on the job that night, sobbing that he wished he had never accompanied Joe Brown that day.

Dan tried to work that afternoon but he was too weak. Buford Dark drove him home, where Margaret dressed the wounds and put Dan to bed. But before Dan dared fall asleep, he warned Margaret that somebody might come back to finish the job, perhaps with a warrant for Dan or Charlie or both.

That night there was unusually heavy traffic lower down in the grove. Margaret could see cars pulling up to the Thomas house, signaling, turning their lights on and off. Remembering that Mrs. Thomas was kin to Turner, she woke Dan up and they discussed whether they were in danger. They decided not to call Sheriff Marshall, Clyde Turner's boss, because they suspected they would get little sympathy. Instead, Dan called the nearest state trooper office, fifty miles away. A trooper visited their home late that night, but the cars left the grove without ever menacing them and the trooper decided there was no danger and left.

Still, the feeling of danger persisted. That night, Charlie York talked about sneaking out of Bradshaw County over the back roads before the mob could get him. The Sizemores had to talk him into staying inside, warning him what might happen if the gang caught him by himself.

On Monday morning, Charlie's money arrived from Detroit and he was free to leave. Dan agreed to drive him down to the Tri-Cities airport. But first Dan called a friend of his who worked with the Virginia Council of Human Relations. The friend made a call to the sheriff of Bradshaw County and notified him that Dan was making a trip to Tri-Cities and was expected to return safe and sound.

After dropping Charlie off at the airport, the Sizemores visited the Tri-Cities office of the FBI that had investigated them when Pete and Chris had gone. An agent promised to investigate this incident, too.

The Sizemores got back home again with no problem. But the feeling of siege continued. The children came home from school and said their own teachers were denouncing the Sizemores and the citizens' council as Communists. Then Dan began getting phone calls from the Bradshaw County sheriff, who requested a private meeting to discuss the situation. Dan warily borrowed a friend's car and drove into Milan by a back road. But when he saw the sheriff waiting with three or four men, Dan suspected a trap and went back home again.

For a solid week, Dan stayed home from work, keeping watch over the house by night, resting during the day. The marks on his face began to recede but his body still ached.

One morning Margaret woke up in a chair, with the hated shotgun in her lap. Seeing Dan sprawled on the couch, she said, "This is enough, Dan. We can't live like this. We've got to get back to normal. Maybe they'll forget about it. But we can't go on like this every night."

Dan went back to work that day. But before he got there, he bought a pistol and slipped it into his pants pocket. It was the first time he had carried a weapon to work since he was a mean young foreman for the McGuires, thirty years before.

Shortly after that, the FBI did make an investigation. They said the beating appeared to be a personal issue, not a matter for them to pursue. The FBI agent said he had received a report from the Montgomerys, next door, accusing Charlie York of being a peeping Tom. The FBI also said the only possible charge they could see was some kind of civil rights violation against Charlie. But since Charlie had gone to the coast, there didn't seem much point in pursuing that.

"I wasn't surprised when they said there was nothing they could do," Dan said. "We had already seen that on the television screen from Chicago in 1968. Mayor Daley's boys could whip people in the park without being prosecuted. Some of my friends who were active in protests in the mountains laughed at me for going to the FBI in the first place. They said I should have known better. I guess I should have. One week after I talked to one friend on the phone, his house was fire-bombed over in Kentucky and the county judge confiscated all his books. They never caught the bomber, but they almost put my friend in jail for conspiracy. So I guess I could have saved myself the trouble. I just went back to work like a good obedient boy."

When Dan returned to the mine, the bruises and cuts were still visible on his face. But his bosses never asked him what had happened. Everybody seemed eager to let the matter drop. By subtle hints, Joe Brown and Webb Harrington let Dan know that the issue was considered closed.

But violence was not finished in the grove. In the summer of 1970, while Margaret and the children were visiting Toronto, Dan noticed Clyde Turner poking around the Montgomery house one afternoon. Dan felt the deputy was not there to assault him, so he watched with curiosity.

The house belonged to Lester Montgomery, a fat, moody young man who now lived by himself since his family moved to Florida. He and his father were active in the local Democratic club in Bradshaw County. Because Dan's citizens' group had usually been active against the incumbent Republicans, Dan had few problems with the Montgomerys.

On this hot summer afternoon, Clyde Turner knocked on the Montgomery door but it seemed locked. Then he broke through the screen and let himself in, only to rush right outside again.

Noticing Dan standing on the porch, Turner said, without any reference to the beating, that Lester Montgomery had not been seen in several days. Turner then asked Dan to help him investigate.

Vaguely fearing some kind of trick, Dan followed the deputy through the opened door. But as soon as he stepped inside, he recognized the most putrid smell he had ever encountered. With their hands over their noses and mouths, they followed the smell until they discovered the bloated body of the fat young man, quivering with a million voracious bugs. A pistol lay on the floor alongside him.

The two men rushed outside, sick to their stomachs, and telephoned for more deputies to come. When the other men arrived, Turner told them that Lester Montgomery had shot himself several days before. The other deputies refused to enter the foul-smelling room, so Turner asked Dan to help carry the body outside. Eager to show his courage to the man who had pistol-whipped him, Dan agreed.

In the sultry summer heat, Dan and Turner dragged Lester's body down a flight of stairs, pausing every few feet to vomit. Outside, with the other deputies keeping their distance,

Dan and Turner placed the body in a sack and removed it in the sheriff's wagon.

At first Dan suspected that he might somehow be set up for blame in the death. Some people whispered that politics might have prompted some kind of revenge killing. But the official verdict never wavered from suicide. Nobody ever knew why.

The sickening memory did not go away easily for Dan. As the heat wave persisted, he lay in his own empty home, unable to eat or move. He tried to pack his lunch to take to the mine but found he could not touch food.

Not until he was deep inside the mine every night did his mind accept the thought of eating. Then, in the man-made cave that had become his natural environment, he would borrow a sandwich from one of his buddies and wolf it down hungrily. Several times during the shift he would repeat this. But at one o'clock in the morning, when he climbed the grove past the Montgomery house, he felt himself growing weak again. The cycle lasted several weeks, until his vacation, when he joined the family in the clean, safe atmosphere of Toronto.

The Montgomery house was soon rented to other families, but they rarely stayed more than a week or two. Rumor said that the foul odor could not be purged from the house. Gradually the house acquired a reputation as a place of bad luck, where families were dissolved by quarrels or ill health. It remained vacant most of the time, an extra buffer between the Sizemores and the community.

The violence seemed to take a vacation in the wintertime but returned in the summer of 1971. Mary-Ann had joined a camping group that toured the country, visiting Indian reservations, meeting migrant workers, camping in rural black communities. The group was mostly rich children from up north, whose parents wanted them to understand other parts of the country. When the tour visited Appalachia, the Sizemores invited them to camp on their front lawn.

The first night was uneventful as the ten campers and two counselors made plans to bunk down for the night. When Dan returned from work at one o'clock, he invited those who were awake to chat with him in the house.

While they were inside talking, they heard a yelling from the lawn. Dan assumed some of the youngsters were playing until he spotted a man running between two tents, carrying a club. Dan reached into his pocket and gripped the pistol he now carried to work. He rushed onto the front porch and fired the pistol twice in the air. He saw several men rushing through the grove, down the hill, toward the main camp. Then all was quiet.

Not knowing who his visitors had been, Dan insisted the campers stay inside that night. He told them he would sit on the front porch with his shotgun.

Around three in the morning, when everybody was asleep, Dan was startled by a shotgun blast, fired from down in the grove. He jumped up, just in time to spot a man crouched in the path. Grabbing his shotgun, Dan fired a shot from the hip. The shot was too low and ricocheted off the wooden floor but apparently it whizzed close enough to the intruder to scare him off.

The next morning, the campers left on schedule, with plenty of good stories to tell about mountain living.

Dan called the sheriff, who said he didn't know anything. But a friend whispered that the intruders had been a gang of ne'er-do-wells from further down in the camp, who had been bothering other families all summer. Apparently the thugs had noticed the carloads of campers and had decided to terrorize them for the sheer sport of it.

Dan realized he knew the gang. They came from families named Horn and Wakefield who traditionally in their early teens dropped out of school and into whiskey and pills and mischief. You could see them by the road, day or night, working on their cars or staring haughtily at anybody who passed.

Dan remembered an incident a few months ago. He had been walking on the main road to catch his ride to work. On the other side of the street, one of the Wakefield boys had spotted Dan and had performed a little dance on one leg, lifting his other leg like a dog about to piss. There was something insolent about the young man that made Dan want to rush across the road and throttle him. But Dan's wisdom got the better of his temper.

The gang had frequently ridden on motorcycles along the strip ledge on Milan Mountain, gunning their motors in the darkness. Virgil Horn was said to have emptied 10,000 gallons of diesel fuel from a strip bulldozer and the company never filed a complaint. Another time Virgil was said to have pushed a stripper's jeep over the side of the ledge. Dan had secretly laughed at the idea of strip miners being victimized. But now the gang was working awful close to home.

The following Saturday night, Dan was rocking on the front porch when he heard a shotgun fired from the strip road. He knew the road was higher than his house. A sniper could probably kill somebody lying in bed, if he really got serious. Once again Dan sat up all night with murder in his heart.

Margaret decided to give the county sheriff another try. But Sheriff Marshall merely pursed his lips and asked, "Have you given any thought to leaving the county?"

This infuriated Margaret, who blurted out that she'd be damned if a bunch of punks could run her out of anywhere.

And the sheriff sighed and said, "Well, Miz Sizemore, it's rough. It's plenty rough."

The thugs finally did kill somebody. A few weeks later, in a drunken brawl, two Wakefield brothers knifed a Horn to death. Denton Wakefield got a five-year sentence, one of the longest murder sentences in the history of Bradshaw County.

Cecil Wakefield left the county and has not returned since. The rest of the bunch went on to other glories elsewhere.

That nearly ended the shooting up in the grove. That November the Sizemores found their pet dog, Blackie, shot to death on the dirt path, halfway down the hill. They never were sure if it was the Baileys who did it. But as Margaret said, "The Sizemores are fair game. It could have been anybody."

25

Margaret has put together Sunday dinner, the only meal the family shares all week. Lily Combs, a young friend of theirs from over in Morgantown, is visiting with her three-year-old son. There is room at the table for her, since David eats in the kitchen by choice and Bobby is down in the valley playing ball. The rest of us cluster into the dining room.

Sunday dinner is pork chops with mashed potatoes and fresh cauliflower and a crisp green salad. Margaret has set out a clean tablecloth, clearing away a week's accumulation of books and papers and games from the table. Dan opens the bottle of cold duck we bought in Morgantown yesterday.

Nobody says grace. Lily starts relating gossip about the Big Ridge executives her family knows. She tells how John Drago, the big boss at the Milan office, is supposed to be retired in

Arizona as soon as the front office in New York gives him the word.

"I guessed as much," Dan sighs. "Old Drago hasn't been keeping his mind on his work lately. He'll get us all killed before he retires in the sunshine of Arizona."

Dan takes a second glass of cold duck. Margaret has not touched her first.

The food is delicious, particularly the cauliflower, boiled until soft, doused with salt and butter. The two little boys finish eating and scramble noisily into the next room. The afternoon sun shines through the closed windows. The room becomes stuffy. We lean back in our chairs while Margaret brings pie and ice cream.

Dan grows weary of the coal gossip. He turns to me and starts talking about the World Series, which has been held since the last time I visited. The Oakland Athletics have just defeated the Cincinnati Reds.

"Well, what do you think of our poor hillbilly Reds?" Dan asks.

Before I have a chance to answer, Dan continues. "I knew they would never hold off the big team from California," Dan says, his tone indicating that he has already formulated his theory and now he wants to try it out on me.

"It was always like that," Dan goes on. "The poor lowly hillbillies from Cincinnati. So intimidated by their miserable state that they choke up against the teams from the coast. Remember how they folded in 1970 against Baltimore? Remember how the Yankees crushed them in 1961? Oh, God, I remember in 1939 when I was just starting out, poor Ernie Lombardi falling in a swoon at home plate while all those Yankees stepped over his prostrate body. Oh, it's awful. But if you live in a hillbilly environment, you expect to fail.

"We used to listen to the Reds' games on the Williamson radio station. They could hold their own against the Saint Louis Cardinals or the Pittsburgh Pirates. But as soon as the

Brooklyn Dodgers came in, forget it. One time the Dodgers scored fifteen runs in the first inning. They would have scored more except they got tired running."

Dan stops for breath. Dan knows I used to cover baseball when I was a sports writer. I decide to give him the benefit of my wisdom. I tell him that Oakland-Cincinnati is hardly a match between the proud West Coast and lowly Appalachia. I tell him about the Oakland A's playing in a dreary stadium built with gray concrete, half-empty on cold, windy nights, when the gusts rattle the beer cans and the scraps of paper fly in the faces of the players.

I try to tell him how San Francisco has given Oakland a massive inferiority complex that is passed along to the players. I try to tell him how the Oakland owner, Charles O. Finley, has made his players insecure by his erratic behavior, how he tried to break Vida Blue's spirit, how he fired managers on whim. Dan merely snorts.

I try to tell him how Cincinnati is the opposite of Oakland, a solid middle-class city with a strong heritage of good schools, good restaurants, good parks, good culture. I try to tell him how Cincinnati players like Pete Rose and Johnny Bench are the most secure, confident athletes in baseball, playing in a warm, civic-minded town. But Dan doesn't want to hear about it. All he wants to know is that Cincinnati collects hillbillies escaping the coal camps.

"You're not from the area," Dan says, his voice rising. "You can't understand what it's like to be a hillbilly. It infects everything you do. Your brain doesn't work as well. You can't perform. You know you're going to fail. You can't compete with outsiders. You know they're smarter than you. Sure, you may whip your buddies down on Shit Creek. But when you go up against the Big City, buddy, you know you're going to get whipped. So what's the sense in trying? That's what Cincinnati's all about."

Dan pours his third glass of cold duck. Margaret still has

not touched her first glass. She looks nervously at Dan. He ignores her. He offers to pour me a second glass but I tell him one is my limit, otherwise I get sleepy.

"I think it would be good if you would have another glass," he says, his eyes flashing behind his thick-rimmed glasses. "It would make you see things more clearly. George, you're a beautiful person but you still haven't learned everything about the mountains. I hope I'm still around in another twenty years. Maybe then you'll come to your senses about those poor dumb mountain ballplayers."

The dinner is over. Margaret and Mary-Ann clear the table while Dan sits on the porch. The sun is starting to set behind the opposite ridge. Even late in October, its rays are warm and bright. Dan rocks on the swing, his arms folded across his chest, his jaw stuck way out.

Bobby and his friends are playing touch football in the front yard, despite the steep angle and the tree stumps and the frequent dips and curves. They recruit me and Vicky and Lily Combs and the two little boys, putting six men on a side—or "persons," if you will. Slowed by all the cauliflower I have eaten, I pay no attention to Lily Combs as she blocks for the opposing team. She whacks me with a forearm in the chest, sending me tumbling on the slanted lawn. Mountain girls don't need women's liberation, at least in touch football.

Dan is attracted by the hard hitting. He ambles down the lawn to watch closer. He pretends to lead cheers, laughing every time somebody drops a pass. But then he trudges up the steps again and lounges on the swing. The game goes on until the gnats begin swarming in the dusk. We go inside for another cup of coffee. Lily collects her little boy and goes home. Margaret goes upstairs to study for an exam tomorrow morning. Dan and I are left sitting in the kitchen. He digs out a jar of amber-colored liquid, unmarked.

"It's home brew," he says, drinking directly from the wide

mouth of the jar. "One of the boys at the mine gave it to me. It's beer. His folks make it. Try it."

Dan offers the jar. The home brew tastes like strong, stale beer, left standing for a week. I pass the jar back to Dan. He takes a sip with studied gusto.

"The ridge people make some of the world's greatest home brew," Dan says. "They talk like ducks, but they know how to make beer."

Sipping steadily from the jar, Dan tell me about the moonshine the men bring to the mine. He tells how they test a new batch by leaving it around for somebody else to sneak a sample first. If the other guy doesn't fall over dead, they know the new batch is fine and they reclaim it. Dan offers me another sip. He asks me why I don't drink more than I do. I tell him I don't like losing control of myself in any way because it spoils my discipline as a person and as a writer. Dan shakes his head. We drop the subject.

In the living room the two younger boys are wrestling, strewing the Sunday newspaper all over the floor. The radio is blasting upstairs. Dan takes a few more sips of home brew and settles back in his favorite chair. We don't say much to each other. The boys finally drift off to bed. Somebody turns off the radio upstairs. Dan falls asleep in his chair.

Monday

26

The house is quiet after the full weekend. David and his stereo are still asleep. Even the fourteen guinea pigs are quiet.

Dan fixes his first cup of coffee amidst the shambles of ashtrays and coffee cups from last night's late snacks. The day is gray and bleak, just about what you can expect from a Monday.

Dan is dressed and shaven, his manner subdued. He stirs his coffee over and over again.

"Margaret says I owe you an apology," he says. "She says I was giving you a terrible rough time of it last night. She says you're such a sweet person that I shouldn't raise my voice like that. I tried to tell her that you understand the ugly parts of me. But she said I should apologize."

I tell Dan to stop poor-mouthing me. That brightens him up.

"Well, I still think you ought to drink more," Dan says.

I mumble something about discipline and control.

"You see!" Dan shouts. "It's just the opposite for me. George, you say you don't want to lose your control over your mind. But I do. Drinking numbs my brain. Drinking helps me forget.

"I don't do it every Sunday, George, you know that, but goddamn it, it's good to get dumbed out once in a while. Look, I make coal mining sound so damn exciting, but let's face it, we're all terrified every time we go over to that place. We never know when the roof is going to fall or something is going to blow up. Look what happened in that smoke on Friday.

"You're living under constant pressure. I feel my lungs starting to give out. Goddamn it, this is no way to live. It's good to forget once in a while. That's why I loved that beautiful dry California white wine you brought last time. My God, that was wonderful.

"Drinking is a mountain man's right, the way I look at it. Margaret doesn't like it but she respects my right to get dumbed out.

"It's rough this morning, though. I've got a foreman's meeting at noon. They'll just bullshit us about safety conditions again and I'll have to sit there and tough it out. When I go in there hung over, I'm as nervous as a whore on the shithouse pot for two-three hours. But you tough it out."

Dan gets up and starts making his lunch. He scrapes together some leftovers from yesterday's meal. His mind is wandering.

"Margaret's doing real good at college," he says. "That's part of our plan, you know. She'd rather study social sciences but she agreed to study nursing, so she'd have a profession in case I got disabled or something. In a couple of years, she'll have her degree as a registered nurse. Then maybe we can pick up and move to Toronto, to be with the boys. I don't think I'd miss Appalachia too much. But right now, for the next couple of years, I guess I'm just going to have to tough it out."

Dan drops half a dozen cookies into his lunch pail. Then he closes the cover and snaps the catch.

Epilogue

A month after this book was completed, Dan Sizemore and all the 175 miners were laid off by Big Ridge. The pink slips in their pay envelopes said that "economic conditions" made it impossible to keep the mine open.

Dan has been unable to find a mining job because his lungs cannot pass the examination. Margaret is still working toward her nursing degree. They are talking of emigrating to Canada.